STATUS AND TRENDS IN SPENT FUEL AND RADIOACTIVE WASTE MANAGEMENT

The following States are Members of the International Atomic Energy Agency:

AFGHANISTAN
ALBANIA
ALGERIA
ANGOLA
ANTIGUA AND BARBUDA
ARGENTINA
ARMENIA
AUSTRALIA
AUSTRIA
AZERBAIJAN
BAHAMAS
BAHRAIN
BANGLADESH
BARBADOS
BELARUS
BELGIUM
BELIZE
BENIN
BOLIVIA, PLURINATIONAL
  STATE OF
BOSNIA AND HERZEGOVINA
BOTSWANA
BRAZIL
BRUNEI DARUSSALAM
BULGARIA
BURKINA FASO
BURUNDI
CAMBODIA
CAMEROON
CANADA
CENTRAL AFRICAN
  REPUBLIC
CHAD
CHILE
CHINA
COLOMBIA
COMOROS
CONGO
COSTA RICA
CÔTE D'IVOIRE
CROATIA
CUBA
CYPRUS
CZECH REPUBLIC
DEMOCRATIC REPUBLIC
  OF THE CONGO
DENMARK
DJIBOUTI
DOMINICA
DOMINICAN REPUBLIC
ECUADOR
EGYPT
EL SALVADOR
ERITREA
ESTONIA
ESWATINI
ETHIOPIA
FIJI
FINLAND
FRANCE
GABON

GEORGIA
GERMANY
GHANA
GREECE
GRENADA
GUATEMALA
GUYANA
HAITI
HOLY SEE
HONDURAS
HUNGARY
ICELAND
INDIA
INDONESIA
IRAN, ISLAMIC REPUBLIC OF
IRAQ
IRELAND
ISRAEL
ITALY
JAMAICA
JAPAN
JORDAN
KAZAKHSTAN
KENYA
KOREA, REPUBLIC OF
KUWAIT
KYRGYZSTAN
LAO PEOPLE'S DEMOCRATIC
  REPUBLIC
LATVIA
LEBANON
LESOTHO
LIBERIA
LIBYA
LIECHTENSTEIN
LITHUANIA
LUXEMBOURG
MADAGASCAR
MALAWI
MALAYSIA
MALI
MALTA
MARSHALL ISLANDS
MAURITANIA
MAURITIUS
MEXICO
MONACO
MONGOLIA
MONTENEGRO
MOROCCO
MOZAMBIQUE
MYANMAR
NAMIBIA
NEPAL
NETHERLANDS
NEW ZEALAND
NICARAGUA
NIGER
NIGERIA
NORTH MACEDONIA
NORWAY

OMAN
PAKISTAN
PALAU
PANAMA
PAPUA NEW GUINEA
PARAGUAY
PERU
PHILIPPINES
POLAND
PORTUGAL
QATAR
REPUBLIC OF MOLDOVA
ROMANIA
RUSSIAN FEDERATION
RWANDA
SAINT LUCIA
SAINT VINCENT AND
  THE GRENADINES
SAMOA
SAN MARINO
SAUDI ARABIA
SENEGAL
SERBIA
SEYCHELLES
SIERRA LEONE
SINGAPORE
SLOVAKIA
SLOVENIA
SOUTH AFRICA
SPAIN
SRI LANKA
SUDAN
SWEDEN
SWITZERLAND
SYRIAN ARAB REPUBLIC
TAJIKISTAN
THAILAND
TOGO
TRINIDAD AND TOBAGO
TUNISIA
TURKEY
TURKMENISTAN
UGANDA
UKRAINE
UNITED ARAB EMIRATES
UNITED KINGDOM OF
  GREAT BRITAIN AND
  NORTHERN IRELAND
UNITED REPUBLIC
  OF TANZANIA
UNITED STATES OF AMERICA
URUGUAY
UZBEKISTAN
VANUATU
VENEZUELA, BOLIVARIAN
  REPUBLIC OF
VIET NAM
YEMEN
ZAMBIA
ZIMBABWE

The Agency's Statute was approved on 23 October 1956 by the Conference on the Statute of the IAEA held at United Nations Headquarters, New York; it entered into force on 29 July 1957. The Headquarters of the Agency are situated in Vienna. Its principal objective is "to accelerate and enlarge the contribution of atomic energy to peace, health and prosperity throughout the world".

IAEA NUCLEAR ENERGY SERIES No. NW-T-1.14 (Rev. 1)

STATUS AND TRENDS IN SPENT FUEL AND RADIOACTIVE WASTE MANAGEMENT

INTERNATIONAL ATOMIC ENERGY AGENCY
VIENNA, 2022

COPYRIGHT NOTICE

© IAEA, 2022

Printed by the IAEA in Austria
January 2022
STI/PUB/1963

IAEA Library Cataloguing in Publication Data

Names: International Atomic Energy Agency.
Title: Status and trends in spent fuel and radioactive waste management / International Atomic Energy Agency.
Description: Vienna : International Atomic Energy Agency, 2022. | Series: IAEA Nuclear Energy Series, ISSN 1995–7807 ; no. NW-T-1.14 (Rev.1) | Includes bibliographical references.
Identifiers: IAEAL 18-01137 | ISBN 978–92–0–130521–3 (paperback: alk. paper) | 978–92–0–130621–0 (pdf) | ISBN 978–92–0–130721–7 (epub)
Subjects: LCSH: Radioactive waste management. | Spent reactor fuels. | Radioactive waste disposal.
Classification: UDC 621.039.7 | STI/PUB/1963

FOREWORD

The IAEA's statutory role is to "seek to accelerate and enlarge the contribution of atomic energy to peace, health and prosperity throughout the world". Among other functions, the IAEA is authorized to "foster the exchange of scientific and technical information on peaceful uses of atomic energy". One way this is achieved is through a range of technical publications including the IAEA Nuclear Energy Series.

The IAEA Nuclear Energy Series comprises publications designed to further the use of nuclear technologies in support of sustainable development, to advance nuclear science and technology, catalyse innovation and build capacity to support the existing and expanded use of nuclear power and nuclear science applications. The publications include information covering all policy, technological and management aspects of the definition and implementation of activities involving the peaceful use of nuclear technology.

The IAEA Safety Standards establish fundamental principles, requirements and recommendations to ensure nuclear safety and serve as a global reference for protecting people and the environment from harmful effects of ionizing radiation.

When IAEA Nuclear Energy Series publications address safety, it is ensured that the IAEA Safety Standards are referred to as the current boundary conditions for the application of nuclear technology.

This publication presents the outcomes of the Status and Trends in Spent Fuel and Radioactive Waste Management project, undertaken by the IAEA in collaboration with the European Commission and the OECD Nuclear Energy Agency. The project was launched in June 2014 and the first cycle was completed in June 2016. The first publication was published in 2018. The second cycle of the project took place from 2016 to 2019, and is an update. One of the aims of the project is to publish regular updates on a three-yearly basis corresponding to the reporting cycle of the Joint Convention on the Safety of Spent Fuel Management and on the Safety of Radioactive Waste Management.

This publication provides an overview of the current status and trends in spent fuel and radioactive waste management, and includes information on current inventories, expected future waste arisings and strategies for the long term management of these materials. The information provided in this publication is based primarily on the national profiles submitted by each of the participating Member States, using a common reference date and data presented in the reports to the Sixth Review Meeting of the Contracting Parties to the Joint Convention. The national profiles are provided on the web site accompanying this publication.

The IAEA is grateful for the participation of all those who contributed to the preparation and drafting of this publication, in particular H. Forsström (Sweden), who chaired the joint working group of representatives from the participating Member States until the beginning of 2018, and E. Neri (Spain), who has been chairing the group since 2018. The IAEA officer responsible for this publication was M. Lust of the Division of Nuclear Fuel Cycle and Waste Technology.

CONTENTS

SUMMARY

Radioactive material is used to treat cancer, monitor the quality of industrial products and generate electricity (among other beneficial uses). In common with all processes, some waste arises from these applications. The waste comprises various forms and materials, with different radioactivity levels and half-lives. Radioactive waste needs to be handled safely and eventually disposed of in a safe manner. Acceptable disposal routes depend on the level of radioactivity and established preferences and practices in different countries. Some waste contains such low levels of radioactivity that it can be released from regulatory control and disposed of as non-radioactive waste. However, for radioactive waste that presents a long term risk to people and the environment, its end point is placement in an appropriate package and disposal in a suitably engineered, multibarrier facility.

Status and Trends in Spent Fuel and Radioactive Waste Management is a collaborative project between the IAEA, the European Commission and the OECD Nuclear Energy Agency, with the participation of nuclear industry organization the World Nuclear Association, that aims to consolidate and complement the information gathered from different initiatives around the world. The objective of the Status and Trends in Spent Fuel and Radioactive Waste Management series is to be the authoritative publication that systematically and periodically summarizes the global status and trends of programmes and inventories for spent fuel and radioactive waste management. The first in the series was published in January 2018 [1], and covered the situation up to the end of December 2013. This is the second edition and covers the situation up to the end of December 2016.

This publication provides an overview of current global inventories of spent fuel and radioactive waste, current arrangements for their management, and future plans for their ultimate disposal where appropriate. Spent fuel is generated only by States operating nuclear power plants or research reactors, whereas radioactive waste is generated in all States producing or using radioactive material in, for example, medicine, industry and research and the nuclear fuel cycle. It is the intention to update this publication at regular intervals, following the reporting schedule for the Joint Convention on the Safety of Spent Fuel Management and on the Safety of Radioactive Waste Management (Joint Convention) [2].

Institutional, organizational and technical aspects of spent fuel and radioactive waste management are explored, including legal and regulatory systems; organization of waste management activities and associated responsibilities; and strategies and plans for ongoing management of different types of spent fuel and radioactive waste, from its generation through conditioning and storage to disposal. This publication compiles the quantities of spent fuel and radioactive waste that currently exist and explores forecasts for the coming decades. Significant trends and the corresponding challenges in the management of spent fuel and radioactive waste are also discussed.

Inventory estimates of spent fuel and radioactive waste in the world are based on information in the National Profiles provided by 38 participating Member States and provided on the web site accompanying this publication. Data are supplemented by published reports to the Joint Convention. For most cases, the information provided corresponds to the end of December 2016; the data are based on information from States accounting for almost 95% of all nuclear power reactors in the world. On this basis, there is an estimated 265 000 tonnes of heavy metal (t HM) of spent fuel in storage worldwide and 127 000 t HM of it has been sent to be reprocessed. The current total global inventory of solid radioactive waste is approximately 38 million m^3, of which 30.5 million m^3 (81% of the total) has been disposed of permanently and a further 7.2 million m^3 (19%) is in storage awaiting final disposal. More than 98% of the volume of solid waste is classified as being very low or low level waste, with most of the remainder being intermediate level waste. In terms of total radioactivity, the situation is fully reversed, with approximately 98% of the radioactivity being associated with intermediate and high level waste. This publication also provides volumes of liquid radioactive waste, both disposed of and in storage.

If naturally occurring radioactive material (NORM) is classified as radioactive waste, depending on the national waste management concept, this is usually considered to be very low level waste (VLLW) or low level waste (LLW). NORM waste is not specifically discussed in this publication, although some countries have reported NORM waste in the National Profiles.

It is evident that significant progress has been made globally in formulating national policies and strategies and in implementing legal and regulatory systems that define responsibilities for the ongoing safe management of spent fuel and radioactive waste. Most States expect to dispose of their waste in facilities located on their territories, with the main focus of international cooperation being on technology development. Disposal facilities for VLLW and LLW are already in operation in several countries. However, in many others, particularly those with small volumes of radioactive waste, disposal options still have to be developed. The most important remaining challenge is the development, public acceptance and long term funding of disposal facilities for high level waste and spent nuclear fuel considered as waste. Significant progress has been made in a few countries, such as the construction licence for a deep geological disposal facility that was granted in Finland in November 2015.

1. INTRODUCTION

1.1. BACKGROUND

Spent fuel and radioactive waste are by-products of the operation of nuclear reactors and related fuel cycle activities, and other uses of radioactive material in medicine, industry and research. Currently two strategies are employed for managing spent fuel from power reactors: either it is considered to be waste or it is considered to be an asset. In the latter case, additional treatment is necessary to recover uranium and plutonium, generating high level waste as a by-product. The radioactive waste comprises various forms and materials, with different radioactivity levels and half-lives.

According to IAEA guidance provided in Safety Standards Series No. GSG-1, Classification of Radioactive Waste [3], the requirements for the management and disposal of radioactive waste are dependent upon its classification: high (HLW), intermediate (ILW), low (LLW) or very low level waste (VLLW). The final disposal of the waste may range from geological disposal for HLW to near surface trench disposal for VLLW. The activity level and the nature of radionuclides in the waste, as well as waste properties, determine the conditioning needs of the waste before disposal or, as the case may be, release from regulatory control.

Status and Trends in Spent Fuel and Radioactive Waste Management (hereafter referred to as the Status and Trends project) is a collaborative project between the IAEA, the European Commission and the OECD Nuclear Energy Agency (OECD/NEA), with the participation of nuclear industry organization the World Nuclear Association (WNA), that aims to consolidate and complement the information gathered from different initiatives around the world. The objective of this series is to provide an authoritative publication that systematically and periodically summarizes the global status and trends of programmes and inventories for spent fuel and radioactive waste management. The first in the series was published in January 2018 [1], and covered the situation up to the end of December 2013. This is the second edition and covers the situation up to the end of December 2016. The analysis presented is based on information as of December 2016 to be consistent with the information in reports presented in 2017 under the framework of the Joint Convention on the Safety of Spent Fuel Management and on the Safety of Radioactive Waste Management (hereafter referred to as the Joint Convention) [2] and those provided to the European Commission in 2018 (if they are published by the Member States) in accordance with Council Directive 2011/70/Euratom of 19 July 2011 establishing a Community framework for the responsible and safe management of spent fuel and radioactive waste (hereafter referred to as the European Atomic Energy Community (Euratom) Waste Directive) [4]. The basic information in this publication has been collected through the submission of National Profiles and has been complemented with openly available Joint Convention National Reports. Approximately 90% of States with operating nuclear power plants submitted National Profiles to the Status and Trends project or their Joint Convention National Reports are openly available, representing almost 95% of all nuclear power reactors in the world. The National Profiles are available on the web site associated with this report.

Some publications cover the subject on a partial or regional basis. The European Commission has published such information every three years since 1992 (see Ref. [5]). The OECD/NEA publishes profiles and reports about radioactive waste management programmes in member countries[1], as well as an annual Nuclear Energy Data report on nuclear power status in NEA member countries and the OECD area [6].

This publication goes further by providing an extensive overview of the management of spent fuel and radioactive waste worldwide and of the quantities involved. The volumes of different types of waste give an indication of the magnitude of the work needed for managing and disposing of the material. A large volume does not, however, necessarily correspond to a large risk or environmental impact. In particular, when determining the potential impact of different materials and waste classes, the radioactivity content needs to be considered. Chemical and other hazardous properties of the waste also need to be considered.

---

[1] See https://www.oecd-nea.org/rwm/profiles.

1.2. OBJECTIVE

The purpose of this publication is to provide a global overview of the status of spent fuel and radioactive waste management programmes, inventories, current practices, technologies and trends. It provides overviews of national arrangements for the management of spent fuel and radioactive waste, of current waste and spent fuel inventories and their future estimates. Achievements, challenges and trends in the management of spent fuel and radioactive waste are also addressed. The data reported are fully dependent on the input from the States and by the assumptions made to transform these data into the waste classes defined by the IAEA Safety Standards Series in GSG-1 [3].

The anticipated audience for the Status and Trends publications include national policy and decision makers and their support staff, as well as professionals in the nuclear and other scientific fields who wish to get an overview of how the spent fuel and radioactive waste is handled in different countries. Although the publication is not directly written for a general audience, much of the information contained herein could be of interest to the public, including the media, researchers, educators and students.

1.3. SCOPE

This publication addresses the following: the institutional, legal and regulatory frameworks for the management of spent fuel and radioactive waste; spent fuel and radioactive waste management programmes, current practices and technologies; and spent fuel and radioactive waste inventories and forecasts. In addition, this publication provides an analysis of the trends and the achievements made in the frame of spent fuel management and radioactive waste management, together with a view on the challenges that are yet to be overcome.

The publication includes all material that a Member State has declared as being radioactive waste, along with spent nuclear fuel (whether the spent fuel has been declared to be waste or not). The collection and compilation of information on spent fuel and radioactive waste reflects the different strategies for spent fuel management (open cycle, closed cycle or awaiting decision) pursued by various countries and involves several challenges, such as the use of different waste classification schemes in different countries, different statuses of waste conditioning and different stages of development of waste management systems.

Other sources of information, in addition to the National Profiles, include the following:

- Openly available National Reports to the Joint Convention, which are produced every three years in advance of the Review Meetings of the Contracting Parties. The IAEA provides the secretariat for the Joint Convention and the latest review meeting was held in May 2018;
- Openly available National Reports to the European Commission in accordance with the Euratom Waste Directive. The latest reporting to the European Commission according to this Directive was due in August 2018;
- Reports to the OECD/NEA providing input to the National Profiles and to other special reports on specific subjects;
- Reports prepared as input to documents by IAEA and OECD/NEA advisory working groups.

The Joint Convention came into force in 2001, with the aim of promoting high levels of safety worldwide in spent fuel and radioactive waste management; as of the end of 2019 there are 82 Contracting Parties to the Joint Convention. The Joint Convention requires that all Contracting Parties prepare, on a triennial basis, a comprehensive National Report describing the measures being taken to implement the obligations of the Joint Convention. These reports are discussed and analysed in the (also triennial) Joint Convention Review Meetings. In the European Union (EU), the Euratom Waste Directive was established in 2011, regarding the responsible and safe management of spent fuel and radioactive waste. The report on the implementation of the Directive is to be submitted by each country to the European Commission every three years. It also requires that each member country of the EU establishes and maintains a long term programme for the implementation of spent fuel and radioactive waste management up to and including disposal.

While national arrangements for ensuring that spent fuel and radioactive waste are safely managed vary from country to country, there are many common features. The arrangements are based on relevant national policies and on corresponding strategies for policy implementation. National policy is typically established at government level, whereas associated strategies are often developed by the organizations with designated responsibility for radioactive waste and/or spent fuel management.

It is evident that significant progress has been made globally in formulating national policies and strategies and in implementing legal and regulatory systems that define responsibilities for the ongoing safe management of spent fuel and radioactive waste. Most countries expect to dispose of their waste in facilities located on their national territories, with the main focus of international cooperation being on technology development.

The majority of the Member States have established a specific waste management organization (WMO) to implement radioactive waste and/or spent fuel management activities, in particular radioactive waste disposal. The WMO may either be a State organization or a private organization. In the latter case, the WMO is typically established by the nuclear power utilities, being the main waste generators, in compliance with requirements established by the government. In countries with limited nuclear activities, the role of the WMO is often taken by the national nuclear research institute or a government ministry responsible for radiation protection.

There are legal frameworks present, so the waste generators are responsible for financing the management and disposal of spent fuel and radioactive waste, or a combination of funding mechanisms exists. As many of these costs manifest a long time after income generation from the activity has ceased (i.e. several decades or more), systems for the advance collection of funds have been established in most countries. In these systems, a special fund is created during the period of operation of the facility; this fund is generally segregated from the main accounts of the relevant business activity, to be used when final decommissioning and waste management activities are performed. A variety of funding methods are currently in use, ranging from funds which are part of the State budget to internal funds in the company generating the waste. In many countries funds are held in separate accounts outside of the waste generating companies or the WMOs, and include State oversight. For legacy waste from early nuclear activities, financing is often provided directly from the annual national budget or through special levies on the relevant industrial sector. Funding of waste management legacies remains a significant challenge in many countries, resulting in delays in the development of disposal facilities, often until long after the waste has been generated. Responsibilities for funding have been established and systems to collect funds have been developed. However, provision of funding for the management of legacies is a significant cause of delay in many countries.

Spent fuel is stored initially in water-filled pools at the reactor site. Water provides both the cooling and shielding necessary for the highly radioactive, fresh spent fuel. As the capacity of these pools is exhausted, storage capacity is typically expanded, either near the respective reactors or at a central location in the country. Both wet and dry storage facilities now exist, with most of the recently constructed facilities being the dry storage type. For the open fuel cycle, the spent fuel will eventually be disposed of in a geological disposal facility following a similar approach as for the HLW.

Although many countries with nuclear power plants have programmes to develop spent fuel or HLW disposal, these programmes are at very different levels of maturity, especially as concerns the siting process and the selection of a site. Three countries, Finland, France and Sweden, have selected a site and are progressing towards licensing and construction. Other countries have time schedules to begin operation of repositories in the 2050s and 2060s and have started an active siting process. The general trend is to site such a facility in a willing and informed volunteer host community.

Disposal facilities for VLLW and LLW are already in operation in several countries, although many others, particularly those with small volumes of radioactive waste, still have to develop their disposal options. The most important remaining challenge is the development of disposal facilities for HLW and/or spent nuclear fuel, where this is considered as waste. In this context public acceptance issues are very important.

For the purposes of this publication, the process for identifying trends was through general discussion among the participating experts, rather than through relative comparison of the quantitative information presented. Some general points may be noted:

- Internationally accepted technical solutions exist to safely and sustainably manage spent fuel and all major types of radioactive waste;
- There has been progress in most countries in the formulation and implementation of national policies, strategies and programmes for the management of spent fuel and radioactive waste;
- Management of spent fuel and radioactive waste is highly regulated;
- The current quantities and future forecasts of spent fuel and radioactive waste are known to a good degree of certainty in most countries;
- There has been significant progress in disposing of radioactive waste and disposed volumes are higher than stored volumes for very low level and low level radioactive waste, which represents most of the volume;
- Spent fuel is safely stored in storage facilities pending implementation of downstream steps, depending on the country's strategy for managing the back end of the fuel cycle — either recycling of fissile materials through reprocessing or direct disposal in a deep geological repository;
- The long term spent fuel and radioactive waste management plans have to address — among other things — knowledge management and preservation, as well as stakeholder involvement (which includes the public). Stakeholder involvement and transparency are important success factors of the implementation of the management of spent fuel and radioactive waste programmes;
- Most countries with a nuclear power programme have dedicated upfront financial systems in place with the purpose of ensuring sufficient and timely financing of the radioactive waste and spent fuel activities. The total costs of management of spent fuel and radioactive waste and decommissioning are relatively modest in relation to the value of the produced energy. However it has to be taken into account that most of the financing is needed once the production of electricity has stopped and there is no more revenue.
- There is active collaboration and cooperation to address spent fuel and radioactive waste management, both nationally and internationally;
- There is a wide range of research and development ongoing to ensure the safe and sustainable management of spent fuel and radioactive waste.

1.4. STRUCTURE

This publication provides an international overview of current spent fuel and radioactive waste management and provides a compilation of policies, strategies and spent fuel and radioactive waste quantities on a regional and global scale. This is based on information in the National Profiles supplied by the Member State or on information provided by the State in its report to the Joint Convention. Section 2 presents the relevant international legal instruments to improve the safety of spent fuel and radioactive waste management. Section 3 outlines the sources of spent fuel and radioactive waste. Sections 4 and 5 explore the frameworks for managing these and give a summary of current strategies, practices and technologies. Section 6 describes the inventories and presents future forecasts. Sections 7 and 8 provide analysis and achievements, as well as trends and challenges, and Section 9 concludes.

Supplemental information is provided in a series of annexes. The structure of the National Profiles and the preparation of this publication were developed by a joint working group on spent fuel and radioactive waste established by the IAEA, the European Commission and the OECD/NEA with the participation of the WNA.

2. INTERNATIONAL LEGAL INSTRUMENTS AND SUPPORTING MATERIALS

There are several international instruments to ensure and improve the safe management of spent fuel and radioactive waste, and to protect people and the environment from the potential for negative effects of ionizing radiation, including international conventions, standards, requirements and peer reviews. These instruments are supported by publications produced by the academic community, the international organizations and organizations involved in the management of spent fuel and radioactive waste, different agreements between organizations and countries, etc. As a result, there are also many scientific and technical reports available, which are available to the community to help improve the management of spent fuel and radioactive waste.

2.1. JOINT CONVENTION ON THE SAFETY OF SPENT FUEL MANAGEMENT AND ON THE SAFETY OF RADIOACTIVE WASTE MANAGEMENT

The Joint Convention [2], which entered into force in 2001, highlights the importance given to spent fuel and radioactive waste management. The Joint Convention applies to spent fuel and radioactive waste created through the processes of civilian nuclear programmes, while spent fuel and other radioactive waste resulting from military or defence programmes fall under the Convention when declared as spent fuel or radioactive waste, or when such materials are transferred permanently to and managed within exclusively civilian programmes. Figure 1 provides an overview of the evolution of the number of Contracting Parties to the Joint Convention over the years. As of March 2020, there were 83 Contracting Parties to the Joint Convention. The Contracting Parties meet every three years to discuss the National Reports, which are subject to a peer review process.[2] There have already been six review meetings, the last of which was held from 21 May to 1 June 2018 [7]. The next review meeting is scheduled in 2021.

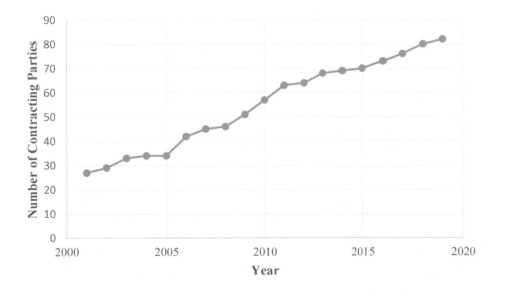

FIG. 1. The evolution of the number of Contracting Parties to the Joint Convention.

---

[2] Information on the Joint Convention [2], its current status, documents and the results of review meetings are available at https://www.iaea.org/topics/nuclear-safety-conventions.

2.2. COUNCIL DIRECTIVE 2011/70/EURATOM OF 19 JULY 2011 ESTABLISHING A COMMUNITY FRAMEWORK FOR THE RESPONSIBLE AND SAFE MANAGEMENT OF SPENT FUEL AND RADIOACTIVE WASTE

The Euratom Waste Directive [4], like the Joint Convention [2], requires appropriate national arrangements for a high level of safety in spent fuel and radioactive waste management. In particular, each EU Member State is required to develop a framework and a programme for the responsible and safe management of spent fuel and radioactive waste, and to implement this programme. This will ensure that an undue burden on future generations is avoided. The Euratom Waste Directive [4] is also intended to ensure adequate public information and participation in the management of spent fuel and radioactive waste. All 27 (28 up until 31 January 2020) EU Member States are members of Euratom and are also Contracting Parties to the Joint Convention.

2.3. INTERNATIONAL SUPPORTING MATERIALS

The Code of Conduct on the Safety and Security of Radioactive Sources [8] aims at helping national authorities to ensure that radioactive sources are used within an appropriate framework of radiation safety and security. The Code is a well accepted, non-legally binding international instrument and has received political support from more than 130 Member States. The Guidance on the Import and Export of Radioactive Sources [9] supplements the Code and aims to provide for an adequate transfer of responsibility when a source is transferred from one State to another. The Guidance on the Management of Disused Radioactive Sources [10] provides further guidance regarding the establishment of a national policy and strategy for the management of disused sources and on the implementation of management options such as recycling and reuse, long term storage pending disposal and return to a supplier.

The IAEA Safety Standards reflect an international consensus on what constitutes a high level of safety for protecting people and the environment from the harmful effects of ionizing radiation. They are issued in the IAEA Safety Standards Series, which has three categories: Safety Fundamentals, Safety Requirements and Safety Guides. IAEA Safety Standards Series No. SF-1, Fundamental Safety Principles, presents the fundamental safety objective and principles of protection and safety, and provides the basis for the Safety Requirements, which establish the requirements to be met to ensure the protection of people and the environment. The Safety Guides provide recommendations and guidance on how to comply with the safety requirements. Of particular relevance for this publication are the following:

— General Safety Requirements GSR Part 3, Radiation Protection and Safety of Radiation Sources: International Basic Safety Standards [11];
— General Safety Requirements GSR Part 5, Predisposal Management of Radioactive Waste [12];
— General Safety Requirements GSR Part 6, Decommissioning of Facilities [13];
— Specific Safety Requirements SSR-4, Safety of Nuclear Fuel Cycle Facilities [14];
— Specific Safety Requirements SSR-5, Disposal of Radioactive Waste [15];
— Specific Safety Guide SSG-1, Borehole Disposal Facilities for Radioactive Waste [16];
— Specific Safety Guide SSG-14, Geological Disposal Facilities for Radioactive Waste [17];
— Specific Safety Guide SSG-15, Storage of Spent Nuclear Fuel [18];
— Specific Safety Guide SSG-29, Near Surface Disposal Facilities for Radioactive Waste [19];
— Specific Safety Guide SSG-40, Predisposal Management of Radioactive Waste from Nuclear Power Plants and Research Reactors [20];
— Specific Safety Guide SSG-41, Predisposal Management of Radioactive Waste from Nuclear Fuel Cycle Facilities [21];
— Specific Safety Guide SSG-42, Safety of Nuclear Fuel Reprocessing Facilities [22];
— Specific Safety Guide SSG-45, Predisposal Management of Radioactive Waste from the Use of Radioactive Material in Medicine, Industry, Agriculture, Research and Education [23];

— General Safety Guide WS-G-6.1, Storage of Radioactive Waste [24].

The IAEA has additionally published many reports related to the safe management of spent fuel and radioactive waste. Some of the related publications from the Nuclear Energy Series are as follows:

— Policies and Strategies for Radioactive Waste Management, Nuclear Energy Series No. NW-G-1.1 [25];
— Options for Management of Spent Nuclear Fuel and Radioactive Waste for Countries Developing New Nuclear Power Programmes, Nuclear Energy Series No. NW-T-1.24 (Rev. 1) [26];
— Management of Disused Sealed Radioactive Sources, Nuclear Energy Series No. NW-T-1.3 [27];
— Framework and Challenges for Initiating Multinational Cooperation for the Development of a Radioactive Waste Repository, Nuclear Energy Series No. NW-T-1.5 [28];
— Storing Spent Fuel Until Transport to Reprocessing or Disposal, Nuclear Energy Series No. NF-T-3.3 [29];
— Stakeholder Involvement Throughout the Life Cycle of Nuclear Facilities, Nuclear Energy Series No. NG-T-1.4 [30];
— Available Reprocessing and Recycling Services for Research Reactor Spent Nuclear Fuel, Nuclear Energy Series No. NW-T-1.11 [31];
— An Overview of Stakeholder Involvement in Decommissioning, Nuclear Energy Series No. NW-T-2.5 [32];
— Locating and Characterizing Disused Sealed Radioactive Sources in Historical Waste, Nuclear Energy Series No. NW-T-1.17 [33];
— Experiences and Lessons Learned Worldwide in the Cleanup and Decommissioning of Nuclear Facilities in the Aftermath of Accidents, Nuclear Energy Series No. NW-T-2.7 [34];
— Communication and Stakeholder Involvement in Environmental Remediation Projects, Nuclear Energy Series No. NW-T-3.5 [35].

Many countries are having peer reviews performed on various aspects of their spent fuel management and radioactive waste management programmes with the aim of assessing and improving their policies and practices. Such reviews are, in fact, required under the Euratom Waste Directive. For transparency, the Member States usually make these peer review reports openly available. The international organizations offer their Member States numerous expert review services related to the peaceful uses of nuclear science and technology. These peer reviews are conducted at the request of member countries and the scope of such reviews is based on the needs of the country.

There is also wide support from academia, the community, national organizations involved in the management of spent fuel and radioactive waste, etc. As a result, there are scientific and technical publications available for the wider community.

3. SOURCES OF SPENT FUEL AND RADIOACTIVE WASTE

All industrial processes result in the generation of waste, which subsequently needs to be managed safely and effectively. The operation of nuclear reactors, as well as their associated fuel cycles (uranium production, enrichment, fuel fabrication and reprocessing), generates radioactive material to be managed as radioactive waste. Radioactive waste also results from the use of radioactive materials in research, medicine, education and industry. This means that all States that engage in any kind of nuclear application have to consider the management of radioactive waste and make sure it is managed in a safe manner, with due regard to the level of radioactivity and in compliance with national regulations, often based on, or in harmony with, IAEA Safety Standards.

Spent nuclear fuel is generated as a result of the operation of all types of nuclear reactors, including power reactors, research reactors, isotope production reactors and propulsion reactors. The spent fuel can be considered to be a resource for reuse or to be waste, depending on the policy and strategy of the Member State.

3.1. RADIOACTIVE WASTE CLASSIFICATION

The activities connected to the safe management of radioactive waste are quite different depending on the type of waste involved. As the radioactivity content of different types of radioactive waste varies greatly, the waste can be assigned to different classes. Waste is classified under national programmes according to their hazards and the available or planned management routes. Although different waste classification systems exist, the classification system used in this publication follows the definitions in para. 2.2 of GSG-1 [3], which classify waste as follows:

(1)   Exempt waste (EW): Waste that meets the criteria for clearance, exemption or exclusion from regulatory control for radiation protection purposes.
(2)   Very short lived waste: Waste that can be stored for decay over a limited period of up to a few years and subsequently cleared from regulatory control according to arrangements approved by the regulatory body, for uncontrolled disposal, use or discharge.
(3)   Very low level waste (VLLW): Waste that does not necessarily meet the criteria of EW, but that does not need a high level of containment and isolation and, therefore, is suitable for disposal in near surface landfill type facilities with limited regulatory control.
(4)   Low level waste (LLW): Waste that is above clearance levels, but with limited amounts of long lived radionuclides. Such waste requires robust isolation and containment for periods of up to a few hundred years and is suitable for disposal in engineered near surface facilities.
(5)   Intermediate level waste (ILW): Waste that, because of its content, particularly of long lived radionuclides, requires a greater degree of containment and isolation than that provided by near surface disposal. However, ILW needs no provision, or only limited provision, for heat dissipation during its storage and disposal.
(6)   High level waste (HLW): Waste with levels of activity concentration high enough to generate significant quantities of heat by the radioactive decay process or waste with large amounts of long lived radionuclides that need to be considered in the design of a disposal facility for such waste. Disposal in deep, stable geological formations, usually several hundred metres or more below the surface, is the generally recognized option for disposal of HLW.

Generally, the higher the hazard, the more elaborated and/or deeper the disposal concept. Depending on national polices, several waste categories are sometimes grouped together for management in a single facility. In this case, the combined facility needs to be designed considering the safety of the highest class of waste it houses. The association between waste classes, activity levels and half-lives, with the boundaries between classes (shown as dashed lines), is illustrated conceptually in Fig. 2.

Some of the radioactive material has such a low content of radionuclides that the radiological impact is negligible, and it can be released from regulatory control ('clearance') in accordance with the State's regulations. This is the case for EW. Other properties, such as chemical hazards, may also affect the available management route. Some countries have a special classification ('mixed waste') that includes non-radiological hazards. An overview of national classification schemes is provided in Annex 1.

Most of the radioactivity associated with radioactive waste is ILW and HLW. While VLLW and LLW comprise more than 90% of the total volume of the waste (see Fig. 3), ILW and HLW typically comprise more than 95% of the total radioactivity.

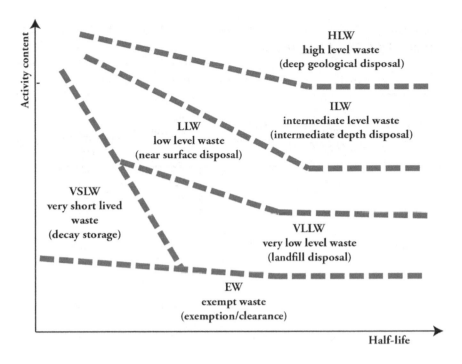

FIG. 2. *Conceptual illustration of the waste classification scheme [1].*

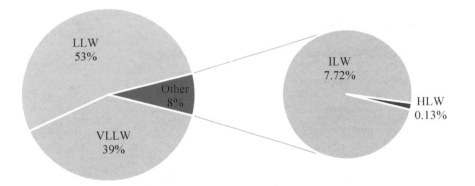

FIG. 3. *Share of different classes of radioactive waste in total volumes in storage and disposal, based on the 2016 inventory data.*

3.2. SPENT FUEL AND RADIOACTIVE WASTE FROM NUCLEAR POWER, RESEARCH AND OTHER REACTORS

In 2016, 448 nuclear power reactors were being operated in 30 countries, generating about 10% of global electricity. Belarus, Turkey and the United Arab Emirates were constructing their first nuclear power plants. Italy, Kazakhstan and Lithuania have shut down their nuclear power reactors, so they do not produce any more nuclear energy (see Table 1). There are States with other types of reactors (e.g. research reactors, isotope production, nuclear powered ships/submarines), which brings the total number of States involved to 58[3].

---

[3] See https://nucleus.iaea.org/RRDB/RR/ReactorSearch.aspx.

TABLE 1. IN OPERATION, UNDER CONSTRUCTION AND DECOMMISSIONING NUCLEAR POWER REACTORS, DECEMBER 2016 [36]

Member State	In operation		Under construction		Decommissioning	
	Number of units	Total net electrical capacity (MW)	Number of units	Total net electrical capacity (MW)	Number of units in decommissioning process	Number of units decommissioned
Argentina	3	1632	1	25	0	0
Armenia	1	375	0	0	1	0
Belarus	0	0	2	2218	0	0
Belgium	7	5913	0	0	1	0
Brazil	2	1884	1	1245	0	0
Bulgaria	2	1926	0	0	4	0
Canada	19	13 554	0	0	3	0
China	36	31 384	21	21 622	0	0
Czech Republic	6	3930	0	0	0	0
Finland	4	2764	1	1600	0	0
France	58	63 130	1	1630	10	0
Germany	8	10 799	0	n.a.[a]	20	3
Hungary	4	1889	0	n.a.[a]	0	0
India	22	6240	5	2990	0	0
Iran, Islamic Republic of	1	915	0	n.a.[a]	0	0
Italy	0	n.a.[a]	0	n.a.[a]	4	0
Japan	42	39 752	2	2653	8	1
Kazakhstan	0	n.a.[a]	0	n.a.[a]	1	0
Korea, Republic of	25	23 077	3	4020	0	0
Lithuania	0	n.a.[a]	0	n.a.[a]	2	0
Mexico	2	1552	0	n.a.[a]	0	0
Netherlands	1	482	0	n.a.[a]	1	0

TABLE 1. IN OPERATION, UNDER CONSTRUCTION AND DECOMMISSIONING NUCLEAR POWER REACTORS, DECEMBER 2016 [36] (cont.)

	In operation		Under construction		Decommissioning	
Member State	Number of units	Total net electrical capacity (MW)	Number of units	Total net electrical capacity (MW)	Number of units in decommissioning process	Number of units decommissioned
Pakistan	4	1005	3	2343	0	0
Romania	2	1300	0	n.a.[a]	0	0
Russian Federation	35	26 111	7	5520	4	0
Slovakia	4	1814	2	880	3	0
Slovenia	1	688	0	n.a.[a]	0	0
South Africa	2	1860	0	n.a.[a]	0	0
Spain	7	7121	0	n.a.[a]	2	0
Sweden	10	9740	0	n.a.[a]	3	0
Switzerland	5	3333	0	n.a.[a]	1	1
Ukraine	15	13 107	2	2 070	0	0
United Arab Emirates	0	n.a.[a]	4	5380	0	0
United Kingdom	15	8918	0	n.a.[a]	26	0
United States of America	99	99 869	4	4468	29	13

Source: Power Reactor Information System (PRIS).
Note: n.a.[a]: not applicable.

3.2.1. Radioactive waste

During the operation of a reactor, different types of radioactive waste are generated. This waste includes filters used in water and air treatment, worn out components and industrial waste that has become contaminated with radioactive substances. This waste has to be conditioned, packaged and stored prior to its disposal. Most of this waste (by volume) has low levels of radioactivity (VLLW or LLW).

At the end of its operating life, a reactor is shut down and eventually dismantled. During dismantling, contaminated and activated components are separated, treated and if necessary managed as radioactive waste. The largest volumes of radioactive waste generated are in the VLLW or LLW classes. Smaller volumes of ILW are also generated. The majority of the waste (by volume) from dismantling is, however, not radioactive and can be handled as industrial waste, in accordance with the country's regulations.

Decommissioning of nuclear reactors and management of decommissioning waste is becoming more and more important, as the current global fleet of power reactors is ageing. There are more than 70 power reactors that have been in operation for more than 40 years, and more than 250 power reactors that have been in use for more than 30 years (see Fig. 4). It is foreseen that an increased number of nuclear reactors will be closed over the next two decades. Furthermore, as shown in Table 1, many nuclear power reactors have been shut down worldwide. However, currently, less than 20 have been completely dismantled. These have given useful experiences of complete decommissioning and handling of the radioactive components as radioactive waste. Several further units are in different stages of decommissioning, ranging from defuelling to actual dismantling. However, there are also reactors that, after removal of the spent fuel, are awaiting future dismantling while being kept under safe conditions. In such cases, decontamination and dismantling might be delayed for up to 50–60 years.

3.2.2. Spent fuel

After its use in a reactor, spent fuel is highly radioactive, emits significant radiation and heat, and is typically transferred to wet storage in a fuel pool for several years. After this period (sometimes referred to as a cooling period), the spent fuel can be safely transferred to storage facilities, either wet or dry, or reprocessing facilities. The length of time that spent fuel stays in various types of storage depends on its characteristics and intended disposition. For example, spent fuel intended to be reprocessed may spend very little time in storage (a few years), while spent fuel intended for direct disposal may spend several decades in storage.

Spent fuel contains uranium, fission products, plutonium and other heavier elements. The exact composition of the spent fuel will depend on the initial fuel type (uranium, thorium, mixed oxide (MOX), etc.) its enrichment (i.e. percentage of fissile content) and the type and operating conditions of the reactor (e.g. thermal or fast neutron spectrum, burnup, etc.). In order to take advantage of the remaining fissile content of the spent fuel, some countries have adopted a closed or partially closed fuel cycle in which the spent fuel is reprocessed, resulting in the extraction and reuse of the uranium and plutonium in new fuel, as well as the separation of waste products (see Section 5.1).

3.3. WASTE FROM NUCLEAR FUEL CYCLE FACILITIES

Figure 5 illustrates an example of possible waste and materials generated at different stages of the nuclear fuel cycle. Since the base material, uranium, is radioactive and new radioactive elements are formed during reactor operation, radioactive waste is generated in all steps of the nuclear fuel cycle. Most of this is VLLW,

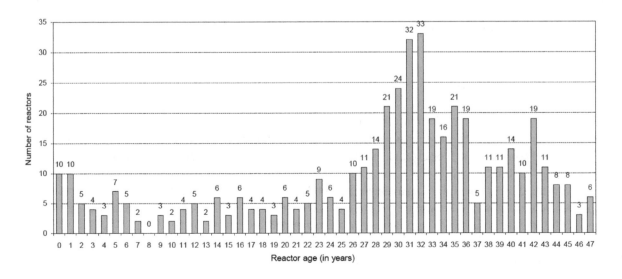

FIG. 4. Global number of operational reactors by age (as of 31 December 2016) [36].

LLW or ILW and is treated according the same principles as waste from nuclear reactors. The exceptions are waste from uranium (or thorium) mining and milling, which is described in more detail in Section 5.5.

Different stages of the nuclear fuel cycle result in radioactive waste and other by-products, as follows:

- Uranium mining and milling (UMM) generates naturally occurring radioactive material (NORM) waste. The waste rock is both the overburden rock, which contains only very low levels of NORM, and the rock from which the uranium bearing material has been separated, which contains residual uranium and other related naturally occurring radionuclides from the uranium decay chain. The mill tailing is the residue after the uranium has been extracted from the uranium bearing material to produce uranium concentrate powder, or so-called 'yellow cake'.
- Conversion of uranium oxide to uranium hexafluoride and back (as part of the enrichment process) generates VLLW and NORM waste.
- Enrichment generates uranium bearing waste (uranium with lower ^{235}U enrichment levels than natural uranium). Depleted uranium (DU) is stored safely (usually in a stable chemical state), although it is not always considered a waste because it can be a resource for MOX fuel, or for down-blending of high enriched uranium to lower enrichments. There are several countries (e.g. France, Russian Federation, United Kingdom (UK)) that require studying the management of DU as waste, if the option of reuse is not implemented on a sufficient scale to use up all of the DU.
- Fuel fabrication generates uranium bearing waste, which is mostly considered to be VLLW. Fuel fabrication from recycled uranium and plutonium may also create alpha bearing waste containing a range of Pu and U isotopes, as well as some of the minor actinides (Am, Np, etc.).
- Reactor operation and maintenance generates a range of waste from VLLW to HLW, mostly waste with activation products created by the neutron bombardment of reactor materials (e.g. ^{60}Co, ^{59}Ni, ^{63}Ni, etc.). However, due to some fuel leakage or due to fissions occurring outside the fuel, this waste can also contain fission products and alpha emitters. The radioactivity circulates in the primary and secondary cooling systems as well as the spent fuel storage pools, and most of it is captured by the cleanup circuits servicing these systems (e.g. filters, ion exchange systems, etc.).
- Water treatment and cleaning processes in spent fuel wet storage facilities may generate filters and resins contaminated with activation products and traces of fission products and alpha emitters.
- In reprocessing facilities, fission products generated in the reactor are extracted from the spent fuel and incorporated into a glass matrix (normally HLW). Claddings and structural components of the fuels are normally considered to be ILW, as well as some technological waste and effluents from the chemical processes. The latter can also be mixed with the HLW in the glass canisters. Reprocessing facilities also generate LLW and VLLW with different radiological contaminants (activation products, fission products and low levels of uranium or plutonium) as part of routine operation and maintenance activities. Manufacturing of new fuel containing recycled uranium or a uranium/plutonium MOX also generates ILW.

3.4. RADIOACTIVE WASTE FROM RESEARCH, MEDICAL AND INDUSTRIAL USE

Radiation can be used to improve quality of life in many ways, and the activities by which radioactive waste is generated include a wide range of activities, including research, the use of radioisotopes as tracers in medical and industrial applications, and the irradiation of materials, such as for sterilization and polymerization. The typical life cycle of the radioactive material is presented in Fig. 6. Generally, the same types of treatment and handling and disposal methods are applied as for similar classes of waste resulting from nuclear power generation.

Sealed radioactive sources are widely used in research, trade, industry, medicine and agriculture. The most common fields of application for radioactive sources in industry include the calibration of measuring devices, materials testing, irradiation and sterilization of products, and level and density measurements. In medicine, radioactive sources are mostly used for radiotherapy and for irradiation of blood. The working

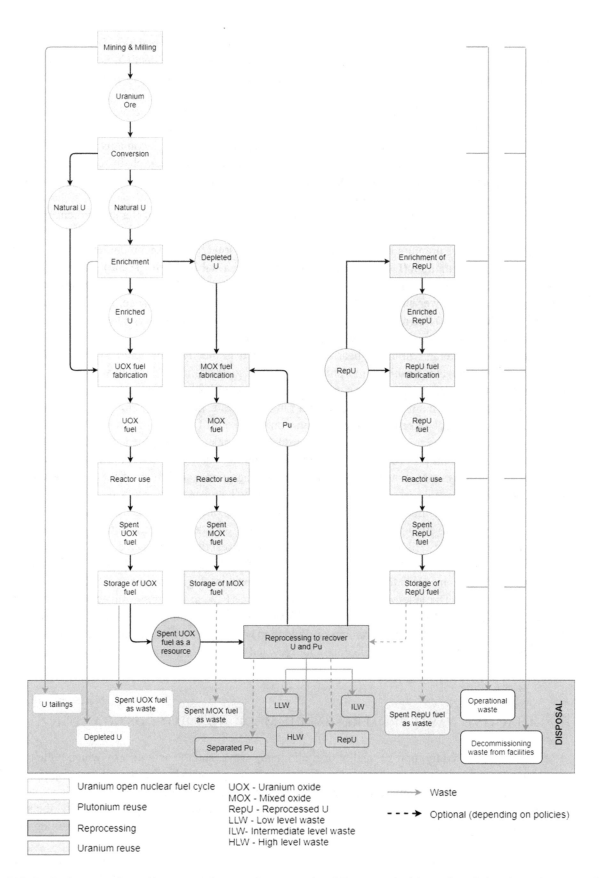

FIG. 5. *A schematic of possible waste and materials generated at different steps of the nuclear fuel cycle in a pressurized water reactor. Different colours indicate different nuclear fuel cycle options. Adapted from [26].*

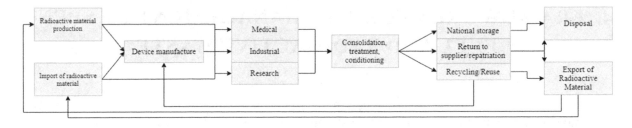

FIG. 6. Life cycle of radioactive material. Adapted from [26].

lives of the sources used vary considerably, on account of the wide range in the half-lives of the radionuclides used. In most countries, devices operated on the basis of a licence for handling are returned to the equipment manufacturer by the operator after end of use, together with the source remaining in the device. The source manufacturer might check for the possibility of further use of the sources and reuse parts of them. Sources that cannot be reused have to be disposed of as radioactive waste. In many countries, disused sealed radioactive sources (DSRSs) are the primary or only type of radioactive waste. A more detailed overview of the management of DSRSs can be found in IAEA Nuclear Energy Series No. NW-T-1.3 [27].

NORM waste can be also produced in different industrial sectors. The most important ones include the phosphate sector, the production of titanium dioxides, water treatment, geothermal energy, steel industry, the oil and gas industry, extraction of rare earths, etc. Depending on the classification of the waste used in the country, NORM might be considered as radioactive waste.

3.5. RADIOACTIVE WASTE FROM MILITARY AND DEFENCE PROGRAMMES

Military and defence activities involving nuclear material create radioactive waste in various forms, and in some cases account for the majority of waste produced in the country. Neither the Joint Convention [2] nor the Euratom Waste Directive [4] requires States to report this waste. However, some States have included military and defence waste in their Joint Convention reports. The aggregated tabulations in Section 6 include any waste being declared and managed as part of the national inventory of radioactive waste.

3.6. OTHER POTENTIAL SOURCES OF RADIOACTIVE WASTE

Other potential sources of radioactive waste include past activities that involved radioactive materials or waste generated by nuclear or industrial accidents. Usually this kind of radioactive waste presents a special challenge, as it may present an additional waste stream where the waste may range from large volume/very low activity to small volume/high activity. Waste forms may be also very variable, so waste management issues may result either from the nature of the radioactive materials (e.g. historical radium bearing waste sites) or from chemical and chemical-toxic aspects. The management needs particular attention if the quantities, location and/or characteristics of the waste exceed the existing waste management infrastructures.

4. FRAMEWORKS FOR THE MANAGEMENT OF SPENT FUEL AND RADIOACTIVE WASTE

National arrangements for securing the safe management of spent fuel and radioactive waste also take into consideration international treaties and standards. A basic prerequisite, as stated in the IAEA's Fundamental Safety Principles [37], the Joint Convention [2] and the Euratom Waste Directive [4], is that the prime responsibility for ensuring the safety of spent fuel and radioactive waste management rests with the licence holder. It is also evident from those documents that the ultimate responsibility for ensuring that programmes are prepared for the management (including disposal) of radioactive waste rests with the State in which that waste arises. These obligations are implemented in each Member State through legislation and regulations in which the roles, responsibilities and reporting relationships of the relevant organizations are established.

4.1. NATIONAL POLICIES

While the national arrangements for ensuring that spent fuel and radioactive waste are safely managed vary from country to country, there are some common features. The national legislative assembly is usually responsible for enacting legislation, which generally includes the establishment of a regulatory body, and in many cases an implementing body for spent fuel and radioactive waste management, as well as defining the essential elements of the national policy and other related governance. Alternatively, national policy can be set out separately by governmental decree or ministerial directives. In some cases, a single policy covering both spent fuel and radioactive waste is adopted, while in other cases, separate policies are issued. The IAEA's guidance on Policies and Strategies for Radioactive Waste Management [25] states that a national policy typically addresses the following:

(a) Responsibilities within the country for spent fuel and radioactive waste management;
(b) Arrangements for financing the management (including disposal and decommissioning);
(c) Preferred management options for spent fuel, policies for waste disposal, import and export of spent fuel and radioactive waste;
(d) Decommissioning of nuclear facilities;
(e) Public information and public involvement in related decisions.

To implement the national policy, one or several strategies have to be developed, which is generally the responsibility of the implementers of waste management practices such as national radioactive WMOs (see Sections 4.4 and 4.5). In some cases, commercial entities and/or agreements with other countries are employed to implement the policy or strategy. Approval of the specific strategy by the regulatory body and/or responsible ministry is also often required.

The usual practice according to the national policy on spent fuel management and radioactive waste management is that final waste is to be disposed of in the country where it is generated. This is also an expectation for Contracting Parties of the Joint Convention, as well a general requirement for EU Member States based on the Euratom Waste Directive [4]. Although spent fuel may be transferred for reprocessing in another country, the HLW or ILW from reprocessing is generally returned to the originating country for long term management.

This does not mean that countries are to be precluded from fulfilling their national obligations through collaboration with other countries [28]. Some countries are seeking joint (or multilateral) solutions for the management of spent fuel and radioactive waste, including disposal in facilities that are operated jointly by, or on behalf of, several countries. The joint/multilateral disposal concept is not be relied on as the only option for radioactive waste management in countries, due to the uncertainties involved.

The export and import of spent fuel and radioactive waste is subject to strict controls. Many States prohibit the import of spent fuel and radioactive waste. Other States, such as France, the Russian Federation and the UK, allow the import of spent fuel from other countries, including those from research and other non-power reactors, for reprocessing services. The current practice is usually to return waste separated from recyclable materials in conditioned form to the country of origin. There are several radioactive waste processing facilities that are used by the waste producers from different countries.

States that are suppliers of sealed radioactive sources for use in medicine and industry, such as Canada, France, Germany, the Russian Federation, South Africa and the United States of America (USA), also accept the return of DSRSs.

4.2. NATIONAL STRATEGIES

Most countries have established national strategies for implementing radioactive waste and spent fuel management, which is in line with Requirement 1 of IAEA Safety Standards Series No. GSR Part 1 (Rev. 1) [38] and stipulated in the Joint Convention [2]. National strategies include, for example, plans for implementing national policy, the development of the required facilities, the identification of roles and the setting of targets for the implementation of the policy. According to the National Profiles on the web site accompanying this publication, many countries have well developed strategies and plans to manage all types of waste, from creation through to final disposal. The slow pace associated with moving towards disposal for ILW, HLW and spent fuel in many countries is dominated by the time required for performing the necessary research and site surveys, engineering, construction and gaining public acceptance of proposals to site facilities in specific areas. For these reasons, some States are progressively implementing their chosen national strategies, especially for the long term management of spent fuel and HLW. An overview of national strategies for spent fuel and different types of radioactive waste is provided in Annex 2.

4.3. LEGAL FRAMEWORK

Requirement 2 of IAEA Safety Standards Series No. GSR Part 1 (Rev. 1), Governmental, Legal and Regulatory Framework for Safety [38], states that "The government shall establish and maintain an appropriate governmental, legal and regulatory framework for safety within which responsibilities are clearly allocated."

This requirement is also reflected in Article 20(2) of the Joint Convention [2]. The legal framework needs to include provisions to ensure sufficient and timely funding of spent fuel and radioactive waste management activities — including providing management facilities and establishing requirements for public involvement in the decision making process. While legal instruments vary, they typically assign roles and responsibilities for nuclear activities, including radioactive waste management, to operating organizations, ministries and other governmental organizations. The National Profiles on the web site accompanying this publication provide information on the national legal frameworks in each of the countries.

4.4. ALLOCATION OF ROLES AND RESPONSIBILITIES

Requirement 2 of GSR Part 1 (Rev. 1) [38] establishes the essential elements of a regulatory framework. At the government level, the ministries or departments of energy, industry, economy and development, with responsibilities for ensuring adequate energy supplies, often support the nuclear power industry in making arrangements for managing spent fuel and radioactive waste. The ministries responsible for ensuring that public health and the environment are adequately protected are typically responsible at the governmental level for issues related to the management of spent fuel and radioactive waste.

The role of the regulator is to ensure that nuclear activities are performed in a safe manner and in accordance with the legal and regulatory framework. For most EU and OECD/NEA members, nuclear safety regulators are now clearly separated from the national ministry in charge of energy or industry [39]. The basic responsibilities for ensuring the safety of spent fuel and radioactive waste management are assigned by all States involved in this study in accordance with the above norms, although in different ways — the differences are usually due to variations in national legislative and regulatory systems. In some countries, for example, the owner or licence holder of a spent fuel and radioactive waste management facility is a private entity and thus is responsible for ensuring safety. In other countries, the owner or licence holder might not be completely distinct from the government and so the responsibility for ensuring the safety of spent fuel and radioactive waste management essentially rests with the State.

4.5. WASTE MANAGEMENT ORGANIZATIONS

Even though the primary responsibility for managing spent fuel and radioactive waste rests with the owner or licence holder of the facility from which the spent fuel and radioactive waste originates, there is a practical need for arrangements at the national level due to the longer term aspects. Many States have created national radioactive WMOs that are responsible for developing arrangements for the disposal of spent fuel and radioactive waste. These WMOs may also be responsible for implementing the management of spent fuel and/or radioactive waste (including HLW from reprocessed spent fuel).

The structure and role of WMOs in countries with a nuclear power programme varies. In some countries, the generators of spent fuel and radioactive waste are responsible for all activities for its safe management, encompassing the disposal of radioactive waste (including HLW from reprocessed spent fuel) and spent fuel. In such cases, the waste generators have formed WMOs that are owned and operated by them. This is true for the management of waste from nuclear power plants in Canada, Finland, Japan and Sweden, for example. In other countries, however, the State has created a separate State-owned organization responsible for waste disposal (including spent fuel and/or all applicable radioactive waste classes), while the responsibility for the interim management of spent fuel and radioactive waste remains with the spent fuel or waste producer. Such an approach is used for managing all radioactive waste in China, France, Germany, the Russian Federation and Switzerland. Other countries may have a mixed approach whereby, for example, private companies are responsible for the short term management of spent fuel, whereas a State-owned or State-controlled body is responsible for the long term management of spent fuel and/or HLW. The private companies might also be responsible in such cases for the management of radioactive waste (with exception of HLW) and/or decommissioning.

However, for countries without nuclear power programmes, the quantity of waste concerned might not justify the existence of a dedicated WMO. In these cases, responsibility for such matters can be taken by a national research centre (e.g. Greece), by a ministerial department (e.g. Luxembourg) or other body [40]. The nature and role of WMOs is given in Annex 3.

4.6. FUNDING ARRANGEMENTS

The Joint Convention [2] requires that a Contracting Party have adequate financial resources available, among other things, to support the safety of facilities for spent fuel and radioactive waste management during their operating lifetime, for decommissioning and also for the activities needed for the operation/closure of disposal facilities. This requires the establishment of a funding system for its spent fuel and radioactive waste management needs, and there are different options available. The country can choose and define the scheme suitable for its particular needs. In most countries, spent fuel and waste producers are responsible for the funding of all activities connected to the management of spent fuel (including direct disposal if it is regarded as waste or disposal of the resulting waste if it is reprocessed) and radioactive waste (including final disposal), and for the decommissioning of the facilities.

An overview of financing schemes and funding mechanisms in different countries is given in Annex 4. The data provided in Annex 4 shows that funding arrangements can sometimes include the costs for management and disposal of all the radioactive waste being generated in a country, while in other cases the funding is limited to the disposal of spent fuel or HLW and the decommissioning of nuclear facilities. In the latter, the costs of management and disposal of other types of waste are paid directly by the waste producers as an operating expense at the time when they occur. The funding arrangements described in Annex 4 mainly relate to spent fuel and radioactive waste from nuclear power plants. In some countries (e.g. Finland and Sweden), this fund also covers the costs of decommissioning the facilities and managing the waste from decommissioning. In other countries (e.g. Switzerland and the USA), separate funds have been established for decommissioning.

For nuclear activities operated by the State, e.g. nuclear research or use of radionuclides in medicine, or for countries having historical or legacy waste dating from past nuclear activities, the State is also most often responsible for the management of the resulting waste. In many cases, the corresponding costs are covered by the State budget (e.g. Latvia). In most cases, no segregated funding arrangement was established. Instead, the funding for current and future waste management is, and will be, met directly from government sources. In some countries, similar arrangements have been implemented for small producers of radioactive waste (i.e. the waste producers pay for waste management and disposal). In other countries, the State takes responsibility for these costs in return for fees paid by the waste producers.

The long time-perspectives associated with radioactive waste and spent fuel management offer challenges to ensure the availability of adequate funds for financing future activities when needed. Some components of importance in this context are the following:

- Funding mechanism;
- Setting fees;
- Calculated and real costs over time and their corresponding uncertainties;
- Costing methodology;
- Management of the funds to ensure their value and growth;
- Safeguarding of the funds against disturbances;
- Ultimate responsibilities for financing.

Most funding systems are based on the premise that the waste producer will pay all costs for the management of the spent fuel and radioactive waste produced and that these costs will be taken from the funds that have been built up. There are different ways to collect funds. In countries with nuclear power reactors the most common method is to levy a fee per kilowatt-hour produced; however, there are also other methods for building the funds, such as establishing a target value of the fund at the end of each year. It is important to keep in mind that many of the costs associated with the management of spent fuel and radioactive waste arise long after the revenue generating activities have ceased. Therefore, it is essential to establish mechanisms to gather and protect funds during the revenue generating phase. For nuclear services provided, e.g. reprocessing or disposal, the funding can be based on a cost per cubic metre of waste delivered or per activity content. Irrespective of which mechanism is used, it is important that the actual fee is based on the best available calculated costs. There are several factors to be taken into account, as follows:

- Expected costs;
- Expected operational lifetime of nuclear power plant and expected future electricity production;
- Expected return on investment of the capital funded;
- Level of security in the funding system.

The typical way of determining the fees is to ensure that the discounted future costs equal the future discounted fee payments plus the present fund content. In the discounting a typical real rate of return (after correction for inflation) is used. The level depends on the expected future development of the country and the expected fund management. Typically, values between 2 and 4% have been used in many countries; however, in Europe, for instance, lower values are currently used, reflecting more pessimistic assumptions on future economic growth. If suitable arrangements are made at an early stage of planning these provisions are relatively straightforward to implement.

For the establishment of fees, the activities to be covered by the funding need to be defined and the expected costs for the necessary facilities and activities calculated. As the timescales are large and many of these costs will occur in the future, it is important to develop a scenario and a time schedule for the management of the spent fuel and radioactive waste. As the cost calculations will inevitably often be based on very early facility designs and operation descriptions, they will involve substantial uncertainties. Additionally, there could also be uncertainties caused by the time schedule. It is also important to address the possible unexpected costs and how to secure the funds for these possible needs. The inclusion of the uncertainties can be handled in different ways, e.g. as a contingency or through a statistical approach. A challenge in the cost calculations is to predict how different cost types will develop in relation to general inflation, and what influence technology development and competitiveness in the industry will have. To ensure that money will be available in the long run, some countries have introduced guarantees in addition to payment of fees. The thinking is that guarantees will cover reduced incomes to the funds, e.g. due to lower power prediction than anticipated or higher than expected cost increases. A guarantee could be in the form of a bank guarantee or a guarantee by the mother company.

The choice of margins included in the calculated costs will be dependent on the safeguards required in the funding system and to what extent the State will take the final responsibility for covering deficiencies in the funding. There might be an extra contingency applied to safeguard against unexpected costs. Alternatively, as is the case for Sweden, the fee would reflect the expected costs, and unexpected costs would be covered by securities.

The collected funds can be managed in different ways. In most cases segregated funds have been established, whereby the management of the funds, i.e. collection, investment and payments, is handled by a dedicated body, often under government control. In other cases, the funds are kept inside the organization and invested, although there are also cases where the funds become part of the State budget. A key question for management of the funds is the effective return on the funds and, in this connection, what investment possibilities exist. To safeguard against cost increases due to inflation and to keep the fees at an appropriately low level it is important that the fund content is invested in such a way that a proper return on the money is achieved. Given that these funds are foreseen to be secured for a long time period, normally the flexibility in investments has been quite low and restricted to very secure investments such as State or property bonds. There are cases where a certain percentage of the funds could be invested in more profitable portfolios, such as shares. The possibilities and restrictions of the investment policy have a strong impact on the return of the funds, but they also influence the stability of the funds and the necessity of liquidity once the use of the funded money gets closer.

As the funds will exist for many decades the risk of disturbances is large. Such disturbances could include, for instance, cost increases, time schedule changes, early reactor shutdown, international and national economic turbulence, bad fund management and companies ceasing to exist. As long as the waste producing activity generates a revenue, it is be possible to adjust the funding requirements through relatively frequent recalculations of the future costs, incomes and returns. This means that changes can be accommodated through a change of the levies on the future waste generation.

In the unlikely case of insufficient funds, for example if the funds are emptied before all activities have been completed, the approaches of each country do differ. In some countries, the State takes over responsibility for covering unfunded costs, while in other countries the waste producer remains responsible for providing additional funding. In the latter case, the risk of insolvency of the waste producer also has to be considered, but in the extreme case it is always the State that takes the final risk. The risk of cost increases due to disturbances will thus be taken on for the future production of nuclear electricity through increased fees and ultimately by the government. Alternatively, one could consider a system such as that in Sweden, where the obligation to pay into the fund remains even after cessation of power production. There are also some cases where the responsibility for an activity and the corresponding funds are transferred from one organization to another, such as when specialist companies take over the dismantling of reactors, as has been done in the USA. Together with the transfer of funds there is also the takeover of all the obligations connected to the activity, including paying all the costs even if they are not covered by the funds.

The total costs of the management of spent fuel and radioactive waste and decommissioning for a mid-size nuclear power programme are relatively modest when compared to the revenue from the electricity generated (around 5–10% of lifetime revenues) [41, 42]. There are both international and national initiatives to estimate the costs, including works by OECD/NEA [42, 43] and the Swedish Nuclear Fuel and Waste Management Company (SKB) [44].

4.7. PLANNING AND INTEGRATION

The ability of a country to deal with its spent nuclear fuel and/or radioactive waste is determined by the extent to which it has established a management policy and strategy, along with supporting infrastructure and legislative and institutional frameworks. The main objective of the management system is to avoid imposing an undue burden on future generations; this means that the generations that produce the spent nuclear fuel and/or radioactive waste have to develop and adequately provide for the implementation of safe, practicable and environmentally acceptable solutions for its long term management. Communication and information sharing between various organizations and various stages is important for minimization of waste generation and effective implementation.

In order to implement an adequate spent fuel or radioactive waste management system, a suitable degree of planning is required, starting with a basic understanding of what types of spent fuel and/or radioactive waste will arise, how much, where and when. The planning needs to be reviewed and updated periodically in order to verify that the planning assumptions are still valid and the overall management plan is still adequate and viable for the country.

Full integration of the spent fuel and/or radioactive waste management system may be challenging, especially in countries with large or complex nuclear industries or ones with a long history of nuclear applications. However, there are usually opportunities to integrate the various agencies responsible for different aspects of the management system, government policy and/or regulations, long-standing practices or infrastructure, etc. A typical practice in the past was to tackle the different types of waste individually. It has to be noted that implemented in this fashion, the lowest cost solution for each individual waste stream may not be the optimal solution for the overall system. By taking advantage of possible synergies between different components of the management system and/or different waste streams, overall optimization may be achieved, leading to a reduction in costs and resource utilization. For example, a national policy could be to send all waste to a single repository, which might be a geological repository suitable for the highest level of waste in the country. If each waste class is destined for a different repository (e.g. engineered landfill for VLLW, engineered near surface for LLW and geological for ILW and HLW), then more work goes into segregating the different classes of waste. The development of alternative routes to disposal requires efficient means in terms of treatment, decontamination and characterization. This issue is important, in particular, for the management of decommissioning waste where large volumes of VLLW or LLW will be generated. In general, a set of evaluation tools will need to be developed to support an integrated supply chain covering all aspects of waste production and management. There have been great successes in some countries, such as in the UK, in diverting a significant part of LLW from the national LLW repository to alternative disposal routes such as licensed industrial landfill. Often the unavailability or the limitation of a disposal option and costs are good drivers to promote the minimization of radioactive waste or recycling of radioactive material, but sometimes, such as in France, the economic trade-off between direct disposal of the waste and waste treatment to reduce disposal volumes or to enable recycling can be difficult.

For countries that are just embarking on a nuclear power programme or national waste management system, there is an opportunity to build in integration right from the start [26]. The availability of existing infrastructure and resources needs to be considered in the planning. This may include collaboration or sharing of services with other countries, especially if there are suitable existing services in one of the countries that can be made available to other countries, rather than each country developing its own infrastructure. This may ultimately lead to the benefit of a much more efficient and cost effective radioactive waste management system in individual countries as well as globally.

Nuclear fuel cycle strategies need to be fully integrated with the overall spent fuel and radioactive waste management policy and infrastructure to ensure an optimal use of resources. Introducing efficiencies into individual steps in isolation can create additional challenges in subsequent steps. One of the main challenges is to maintain enough flexibility to accommodate the range of potential future options for the management of spent fuel, as well as to define and address the relevant issues in storage and transportation.

4.8. MINIMIZATION IN THE MANAGEMENT OF RADIOACTIVE WASTE

Minimization of radioactive waste is the process of reducing the amount and activity of radioactive waste to a level as low as reasonably achievable (ALARA). This is important at all stages, from the design of a facility or activity to decommissioning, and is achieved by design and operations to reduce the amount of waste generated by means such as recycling and reuse, and by treatment to reduce the waste's activity, with due consideration for secondary waste as well as primary waste. Minimization principles have to already have been followed in the planning stages to achieve the best results.

As radioactive waste is defined in the IAEA Safety Glossary [45] as material for which no further use is foreseen and that contains, or is contaminated with, radionuclides at activity concentrations greater than clearance levels as established by the regulatory body, it will be difficult to redefine the material once it has been declared to be radioactive waste. The recycling of radioactive material is a process whereby material is converted into new products, so there is a reduction of wastage of useful materials, use of raw materials and energy use.

There can be limited possibilities for recycling or reuse of radioactive material, especially if it is supposed to be done nationally. Reuse of radioactive material means that an item will be used again, either to perform the same function or a different function. It needs be noted that there are different national approaches, depending on how the radioactive waste is defined. For example, in some countries, radioactive waste is considered only as the material is going to disposal. There are several other criteria to consider in minimization, such as radiation protection and public acceptance. These can weaken the advantage of waste minimization because of doses to workers during waste processing. The opportunity for recycling of materials from a nuclear facility can also be restricted or reduced because of opposition from the public.

This means that for minimization in different phases of spent fuel and radioactive waste management there are different factors that need to be considered, as follows:

- Regulatory and licensing issues: compliance of the option with the applicable regulation;
- Technical and operational issues: availability of technology and facilities to process waste;
- Safety and ALARA issues;
- Economic and schedule issues: costs, duration of implementation of solutions, compatibility with agenda of waste generation;
- Public acceptance and stakeholder issues.

4.9. STAKEHOLDER INVOLVEMENT

Stakeholder involvement, which is an integral part of a stepwise process of decision making, may take the form of sharing information, consulting, dialoguing or deliberating on decisions at different phases. It has to always be seen as a meaningful part of formulating and implementing good policy.

It is important to secure stakeholder involvement through the life cycle of all nuclear facilities, including spent fuel storage facilities and final radioactive waste disposal. International experience has shown that, especially in the case of disposal facilities, the project's progress often relies upon public support. Decision making on long term spent fuel and radioactive waste management is complex, as it not only concerns the current generation, but also future ones, since disposal facilities are designed to operate for many decades and to contain the hazard for thousands of years.

The stakeholders' expectations have to be taken into consideration through different activities and interactions in order to enhance the satisfaction of interested parties [30]. It can be useful and helpful to involve the community early in the decision making process. This helps to build mutual trust between operators, government authorities and stakeholders, especially among the general public. Increased public participation in decisions can promote a greater degree of understanding of the issues related to nuclear power and spent fuel/radioactive waste management, especially with regard to actual risks and benefits. Public confidence is improved when issues that are raised by the public are taken seriously and are carefully and openly evaluated [46].

There are several possibilities for the involvement of interested communities in the siting and development of spent fuel and radioactive waste management facilities. Many siting programmes now incorporate local partnerships and there are even examples of waste management implementing bodies proposing to involve local stakeholders in joint studies, and in the interpretation and review of ongoing site investigation, assessment of the potential impacts on human health and the environment, and development of plans for monitoring these issues during facility operation and closure [47].

The typical steps for implementing stakeholder involvement programmes [30] can be listed as follows:

- Develop a strategy for stakeholder involvement;
- Develop plans for implementing this strategy;
- Ensure that the capacity to effectively implement these plans is available;
- Implement these plans;
- Continually monitor the effectiveness of these actions and look for ways to improve.

In order to earn trust with stakeholders, it is also important to make sure that the licensing authorities are competent, as well as independent from political and industrial influence in their decision making and deliberation. Already many regulators incorporate public comment sessions in their licensing and review processes. It is also crucial to establish clear criteria for the decision making process, so it is understandable how and when, for example, the facility siting process can move from one step to the next. There needs to be clarity on the scope for decision making and identification of the point in the process when specific decisions are finalized and not subject to being revisited [48].

Although decision making processes vary considerably by Member State, depending on culture, history and governmental structure, stakeholder involvement is worthy of consideration. Stakeholder involvement is an essential component of various international conventions and treaties, most commonly related to the strategic environmental assessment and environmental impact assessment. The Convention on Access to Information, Public Participation in Decision-Making and Access to Justice in Environmental Matters, otherwise known as the Aarhus Convention [49], has more than 40 Contracting Parties and is not only an environmental agreement, but also covers also government accountability, transparency and responsiveness. It grants the public rights and access to information. The Convention on Environmental Impact Assessment in a Transboundary Context, otherwise known as the Espoo Convention [50], has more than 40 Contracting Parties, and sets out the obligations of the Parties to assess the environmental impact of certain activities at an early stage of planning and make sure that the stakeholders are involved in the process. It also lays down the general obligation of States to notify and consult each other on all major projects under consideration that are likely to have a significant adverse environmental impact across boundaries. At the same time the Euratom Waste Directive [4] ensures the provision of necessary public information and participation in relation to spent fuel and radioactive waste management while having due regard to security and proprietary information issues.

5. SUMMARY OF CURRENT STRATEGIES, PRACTICES AND TECHNOLOGIES

The previous sections covered the sources of spent fuel and radioactive waste, as well as the frameworks for their safe management. This section gives an overview of the current strategies, practices and technologies for safe management of spent nuclear fuel and radioactive waste. Other properties, such as chemical hazards, may also affect the available management route.

The preferred strategy for the management of all radioactive waste is to contain it (i.e. to confine the radionuclides to within the waste matrix, the packaging and the disposal facility) and to isolate it from the accessible biosphere [15]. Disposal is defined as intentional emplacement in a facility without the intent to retrieve. Disposal options are designed to contain the waste by means of passive engineered and natural features and to isolate it from the accessible biosphere to the extent necessitated by the associated hazard. Figure 7 illustrates the disposal options based on the classes of radioactive waste.

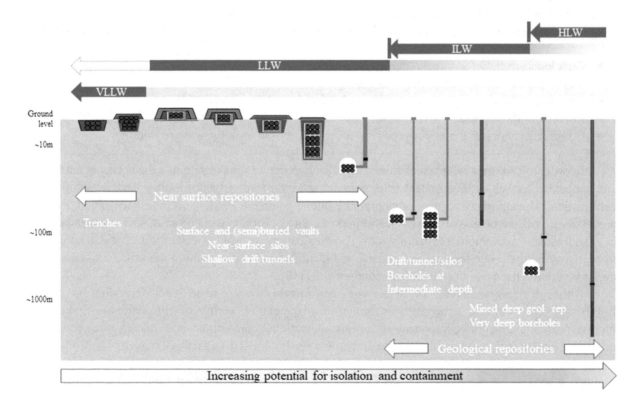

FIG. 7. Conceptual illustration of disposal concepts for different classes of radioactive waste.

5.1. SPENT FUEL AND HIGH LEVEL WASTE

Spent fuel is generated from the operation of nuclear reactors of all types, including research, isotope production, power production, district heating and propulsion reactors. By volume, HLW forms less than 1% of the global volume of radioactive waste, but it consists of about 95% of the total activity of the radioactive waste. The activity level of HLW is high enough that heat generation has to be considered in the design of the waste management facilities. In countries where spent fuel is considered to be waste, it is classified as HLW.

5.1.1. The 'open cycle' and 'closed cycle'

The currently envisaged strategies to ensure a safe and cost effective overall management of spent fuel differ from one country to another and can be described as follows:

- The 'open cycle', 'once through' or 'direct disposal' strategy, in which spent fuel is considered as waste;
- The 'closed cycle' (including the 'partially closed cycle') strategy, in which spent fuel is considered to provide a potential future energy resource.

In the open cycle option, spent fuel is stored for several decades to allow the decay heat to be reduced. After a period of storage, the spent fuel will be encapsulated in a robust, corrosion resistant container to meet disposal acceptance criteria and will be disposed of in a deep geological repository (DGR).

In the closed cycle the spent fuel is reprocessed in order to recover valuable fissile materials (uranium and plutonium). In reprocessing spent fuel is separated into several main components: uranium, plutonium and HLW (containing minor actinides, fission and activation products). HLW (along with other waste such as LLW and ILW) resulting from reprocessing is then stored to allow the decay heat to be reduced pending future disposal, normally in a DGR. The uranium and plutonium can be recycled as nuclear fuel for reactors, while the minor actinides, fission and activation products are currently considered to be waste products. The minor actinide and fission product waste, as well as the main activation products (hulls and end-pieces) from reprocessing can be conditioned in a stable matrix, compacted or vitrified and stored in a very stable matrix purposely designed for storage, transport and disposal. The spent fuel might go through one or more cycles of reprocessing in order to recover valuable material.

Currently, the countries that operate large scale reprocessing facilities are France, India and the Russian Federation. The UK formerly operated two reprocessing facilities. In 2018, the UK completed its reprocessing commitments to its domestic and overseas customers for the management of advanced gas cooled reactor and light water reactor (LWR) spent fuels, and therefore ceased commercial operations at its Thermal Oxide Reprocessing Plant (THORP) reprocessing facility [51]. The strategy for managing Magnox spent fuel is reprocessing and operations are expected to complete in 2021. China is operating a pilot plant and is looking to deploy an industrial facility. Japan is planning to commission its Rokkasho-mura plant by 2021. Other countries, including Belgium, Bulgaria, Hungary, Ukraine, the Czech Republic, Germany, Italy, Japan, the Netherlands, Slovakia, Spain, Sweden and Switzerland, have used services provided by foreign facilities (in the UK, France and the Russian Federation (including during the time of the former USSR)) for the reprocessing of their spent fuel.

The commercial capacity for reprocessing was 4400 tonnes of heavy metal per annum in December 2016 (see Table 2 in grey). The end of reprocessing operations in the UK by the end of 2021 will reduce the worldwide availability of reprocessing capacity until new facilities in the Russian Federation, China and Japan come into operation.

Spent fuel reprocessing in another country is subject to strict controls and is performed on the basis of commercial contracts under the umbrella of bilateral national agreements. In most cases, these commercial contracts provide that the valuable fissile material (usually in the form of fuel for recycling), together with the conditioned HLW from spent fuel reprocessing (as well as fuel component compacted waste in some cases), are sent back to the country from where the spent fuel originated.

TABLE 2. COMMERCIAL SCALE REPROCESSING FACILITIES (DECEMBER 2016)

Country	Facility	Capacity (t HM/a)	Status
France	UP2-400, La Hague	400	Under decommissioning
	UP2-800, La Hague	800	In operation
	UP3, La Hague	800	In operation
	UP1, Marcoule	600	Under decommissioning
Japan	Rokkasho-mura	800	In commissioning
Russian Federation	RT-1, Mayak	400	In operation
	RT-2, Zheleznogorsk	60	Under construction
UK	NDA THORP, Sellafield	900	Ceased operation in 2018
	NDA Magnox Reprocessing, Sellafield	1500	In operation until 2021
Total for facilities operational in December 2016 (in grey)		4400	

Source: IAEA Integrated Nuclear Fuel Cycle Information System.

Reprocessing spent fuel using aqueous separations results in a high level liquid waste that is typically vitrified, i.e. conditioned to produce a chemically durable and heat and radiation resistant engineered solid matrix waste form. Several types of glass (e.g. borosilicate and phosphate) and some ceramics are used for the treatment and conditioning of HLW. The glass containing waste is discharged into containers, which can also be used for storage. The vitrification process and all container handling operations are performed remotely in shielded cells. Significant experience has been obtained with the vitrification process in Belgium, France, Japan, the Russian Federation, the UK and the USA. The HLW is then stored for some decades to allow levels of heat generation to be reduced, in a similar way as for spent fuel. Following storage, HLW, like spent fuel that has been designated as waste, is to be disposed of in a DGR (see Section 5.1.4). Some countries include fuel cladding and structural material that was separated during reprocessing within the HLW class.

The uranium separated during reprocessing, the so-called RepU, can be recycled as fuel in present-day reactors following conversion and re-enrichment if necessary. Recycling RepU from the reprocessing of LWR fuel in a pressurized heavy-water reactor, such as a Canada deuterium uranium (CANDU) reactor, is also developed in China. As an example, natural uranium equivalent fuel is an innovative fuel designed to work in synergy with current and planned spent fuel reprocessing technologies in China. It blends RepU from LWRs with DU to create natural uranium equivalent fuel powder that is used to fabricate pressurized heavy-water reactor fuel [52].

The separated plutonium can be recycled into MOX fuel, in which DU and plutonium oxides are combined. MOX fuel has been used for decades in LWRs worldwide and in a few Generation IV reactors (France in the past, and today in the Russian Federation), where the energy value of the uranium and plutonium can be better utilized. At present only one country, France, has a commercial MOX fuel fabrication facility in operation for manufacturing of LWR fuel. This facility, MELOX, has provided MOX fabrication services since 1995 for France and several other countries [53]. Belgium and the UK also operated MOX facilities (respectively called Belgonucléaire and SMP), which ceased operation in 2006 and 2011. MOX plants are also planned to come into operation over the next few years, as is the case in Japan. Additional multirecycling options in LWRs are under development, such as the regenerating mixture (REMIX) fuel currently in the demonstration phase in the Russian Federation (recycling all the uranium and plutonium without separating them and topping up with some fresh uranium enriched to a higher level than 5%), as well as some other fuel concepts under study in France (MIX and Corail).

There are ongoing initiatives to use DU as fuel or to recycle all recovered long lived actinides together (i.e. with plutonium) in fast reactors. This strategy would make it possible to increase utilization of the uranium in nuclear fuel from less than 1% to well over 90%, which would result in waste containing mainly short lived fission products, thus reducing the waste disposal burden. New reprocessing technologies are being developed to be deployed in conjunction with fast neutron reactors that will burn all long lived actinides, including all uranium and plutonium, without separating them from one another. Several countries implementing or considering reprocessing today have plans, at different stages of development, for future fast breeder reactors, though at present only the Russian Federation operates such reactors at a commercial scale (BN-600 and BN-800).

A summary of the fuel cycle strategies adopted in different countries is given in Table 3. It shows that a majority of countries have adopted or use for referencing the open cycle, while the countries with some of the largest nuclear programmes, e.g. France, the Russian Federation, Japan, India and China, have adopted the closed cycle. Some countries with a small nuclear fleet, like the Netherlands, have also opted for the closed cycle strategy, with reprocessing services provided by one or more of the larger countries with this capability. Table 3 shows also that although several countries have chosen open or closed cycle, there are also countries that are keeping their options open.

TABLE 3. NUCLEAR POWER FUEL CYCLE STRATEGIES

Country	Commercial scale reprocessing facility		Spent fuel currently in another country for reprocessing	Earlier reprocessing, but practice currently ceased	Planning direct placement of spent fuel in a repository	Keeping options open
	Existing	Planned				
Argentina						✓
Belgium[a]				✓	✓	✓
Brazil						✓
Bulgaria[a]			✓			
Canada					✓	
China[b]		✓			✓	
Czech Republic[a]				✓	✓	
Finland				✓	✓	
France	✓					
Germany				✓	✓	
Hungary[a,c]				✓	✓	
India	✓	✓				

TABLE 3. NUCLEAR POWER FUEL CYCLE STRATEGIES (cont.)

Country	Commercial scale reprocessing facility		Spent fuel currently in another country for reprocessing	Earlier reprocessing, but practice currently ceased	Planning direct placement of spent fuel in a repository	Keeping options open
	Existing	Planned				
Italy			✓			
Japan[d]		✓	✓			
Korea, Republic of						✓
Lithuania					✓	
Mexico						✓
Netherlands			✓			
Romania					✓	
Russian Federation	✓	✓				
Slovakia				✓	✓	
Slovenia					✓	
Spain				✓	✓	
Sweden				✓	✓	
Switzerland				✓	✓	
Turkey					✓	
UK[e]	✓			✓	✓	
Ukraine[f]			✓	✓		✓
USA				✓	✓	

[a]   Mixed policy: some fuel has been or will be reprocessed; other fuel will or may be direct disposed.

[b]   The main policy in China is domestic reprocessing. However, some fuel, mainly from CANDU reactors, is planned for direct disposal.

[c]   Earlier fuel returns to the Russian Federation, but no requirement to return waste from reprocessing to Hungary.

[d]   Commercial scale facility at Rokkasho-mura has been constructed and is undergoing test operation.

[e]   The UK has ceased reprocessing on expiry of current contracts.

[f]   Some spent fuel is sent to the Russian Federation for reprocessing. Other fuel is stored awaiting a final decision.

5.1.2. Transport of spent fuel and high level waste

The management of spent fuel and HLW involves a number of transport steps between nuclear power plants, storage facilities, encapsulation/packaging facilities and/or reprocessing facilities, as well as eventually to disposal facilities [54]. Most transport operations are performed within one country, but some journeys cross national frontiers. For countries reprocessing their spent fuel but having no reprocessing facilities of their own, such transboundary movements are necessary. Similarly, the transboundary movement of spent fuel is necessary for countries sending spent fuel from research reactors and other reactors back to the country of origin of the fuel.

Transport is typically undertaken in specially designed transport containers that provide security, shield workers and the general public, and perform other nuclear safety functions such as managing decay heat, ensuring subcriticality and providing neutron shielding [55]. These transport operations are strictly controlled according to national regulations, which are often based on the transport regulations in IAEA Safety Standards Series No. SSR-6 (Rev. 1), Regulations for the Safe Transport of Radioactive Material (2018 Edition) [56]. Each State involved in a transboundary movement has to take the appropriate steps to ensure that the transport operation is undertaken in an appropriate manner and with the authorization of the countries of origin, destination and transit.

5.1.3. Storage

After spent fuel has been discharged from the reactor, it is usually stored for some time in a water-filled spent fuel pool to cool it and provide shielding from its radioactivity. The length of the storage period varies from a few years up to several decades, depending on the spent fuel management strategy adopted. Usually when spent fuel is recycled, the storage period is generally relatively short — a decade or less. In countries that have decided on a direct disposal option or that have yet to make a decision, the storage period can be much longer. Storage systems include wet storage in storage pools or dry storage in storage casks, canisters or vaults built for the purpose [29, 57].

All nuclear power reactors have spent fuel storage pools for the initial decay heat cooling storage period upon discharge from the reactors. They were included in the original design of the reactors. Additional storage capacity, wet or dry, can be built to provide additional storage capacity as needed. The new storage facilities are built outside the containment building, known as away-from-reactor (AFR) stores, and can be either inside or outside the boundaries of the nuclear power plant.

Access to an AFR site may require transport over public roads, railways, sea lanes, etc. AFR facilities are typically purpose built, under a separate licence, for spent fuel storage located away from the main reactor buildings or site. They can be dedicated to one or multiple reactors or they can be a centralized facility serving more than one nuclear power plant. The storage technology for new AFR stores was initially wet storage (see Fig. 8), but dry storage techniques of different types have been developed (see Fig. 9) and are now widely adopted. Examples of existing AFR spent fuel stores (both on reactor sites and outside reactor site boundaries) are given in Table 4. Reprocessing facilities are normally equipped with large AFR pools at the reception for buffer storage before reprocessing.

The canisters for HLW (produced during reprocessing of spent fuel) are stored in air cooled vaults or casks similar to those used for spent fuel storage. Each reprocessing plant has large vaults for canister storage — mainly for its national HLW. In Germany and Switzerland, HLW is stored in casks, while Belgium, Japan and the Netherlands use dry vault storage technology, e.g. the HABOG facility in the Netherlands [58].

5.1.4. Disposal

There is a broad consensus among technical experts that the preferred method of ensuring long term safety for spent fuel and HLW is isolation in a DGR. Geological disposal facilities for long lived waste will provide passive multibarrier isolation of radioactive materials. Emplacement in carefully engineered structures buried deep within suitable geological formations provides the long term stability typical of a

FIG. 8. The wet AFR storage of spent fuel at the Central Interim Storage Facility for Spent Nuclear Fuel (CLAB), Sweden (courtesy of SKB).

FIG. 9. The dry storage hall at Zwilag Zwischenlager Würenlingen AG (courtesy of Zwilag).

TABLE 4. EXTENDED AT-REACTOR AND AFR STORAGE FOR SPENT FUEL, AS AT END OF 2016

Country	Spent fuel storage type
Argentina	Wet AFR
Belgium	Wet store and dry cask at nuclear power plants (NPPs) (depending on plant)
Brazil	Dry cask storage at NPPs
Bulgaria	Wet AFR in use, dry AFR in construction
Canada	Dry cask or modular vault storage at each NPP
Czech Republic	Dry cask storage at NPP
Finland	Wet stores at NPP
France	Wet stores at NPPs, wet store at reprocessing plants before reprocessing
Germany	Dry cask storage at NPPs and there is one central wet storage facility at the Obrigheim NPP
Hungary	Wet storage at NPP, modular vault dry storage facility in the vicinity of Paks NPP
Japan	Wet store and dry cask storage at NPPs and central dry cask storage, wet storage at reprocessing plants before reprocessing
Korea, Republic of	Dry cask storage at one NPP
Lithuania	Dry cask storage at Ignalina NPP
Netherlands	Wet storage at NPPs before transport to France to reprocessing
	Wet storage at reprocessing facilities before reprocessing
Russian Federation	Central wet and dry storage vaults are available for the high-power channel-type reactor (RBMK) and water cooled, water moderated power reactor (WWER) fuel at the Mining and Chemical Complex in Krasnoyarsk
Slovakia	Wet central AFR at one NPP
Spain	Dry cask storage at NPPs
	Central dry vault storage under licensing at Villar de Cañas
Sweden	Central wet storage facility at one NPP
Switzerland	Wet pool storage (Goesgen) and dry cask storage (Beznau) at NPPs and dry cask storage at centralized facility (ZWILAG)
Ukraine	Wet pool storage (ISF-1) at Chornobyl NPP and dry storage facility (DSFSF) at Zaporizhzhya NPP
	Dry storage facility (ISF-2) at Chornobyl NPP and centralized storage facility for WWER spent nuclear fuel (CSFSF) are under construction

TABLE 4. EXTENDED AT-REACTOR AND AFR STORAGE FOR SPENT FUEL, AS AT END OF 2016 (cont.)

Country	Spent fuel storage type
UK	Wet storage at centralized storage both prior to reprocessing and for storage pending potential disposal to a DGR Dry cask storage at one NPP
USA	Dry cask storage at NPPs and AFR sites, one AFR wet pool, proposed centralized dry cask storage under consideration at two sites

stable geological environment [59, 60]. Countries that need to dispose of their spent fuel and HLW are studying different available geological media for it [61].

In the case of the open cycle option, before being sent to the DGR, the spent fuel will have to be encapsulated in a corrosion resistant and mechanically stable container, which will provide isolation for a suitable duration (often thousands of years or more). The vitrified HLW waste form in a stainless-steel canister is specifically designed for long term durability in storage and disposal. In some countries an additional corrosion resistant overpack is also considered. The requirements for container life and integrity depend on the DGR concept and the chosen geological medium.

Posiva Oy in Finland received a construction licence from its regulatory authority in 2015 and construction is ongoing. The operation licence application will be submitted in 2020. The final disposal is scheduled to start in the 2020s. Figure 10 provides an example of the spent fuel canisters to be used in DGR in Finland. In 2011 SKB submitted an application for a spent fuel repository to the regulatory authorities in Sweden. The main hearing was held in the Land and Environment Court in 2017, and in 2019 SKB submitted the documentation required for the government to make the decision. The design process for Cigéo (the Industrial Centre for Geological Disposal) is continuing in France, with a licence application to the regulator anticipated in 2020. There are formal site selection processes under way in several other countries, such as Canada, Germany and the UK [61]. In countries with both spent nuclear fuel and HLW for disposal, a single DGR for both materials is a typically adopted approach. Most other countries with spent nuclear fuel are working towards national solutions, although they are mostly at the early planning stage. Some countries have also indicated an interest in developing multinational disposal facilities, in addition to their own national programme.

Research related to DGR options has been undertaken for several decades using a range of underground research laboratories (URLs). These URLs have an important role in waste disposal programmes and are also valuable in building confidence in national programmes [62, 63]. Currently there are about 20 URLs in use, for example HADES in Belgium, KURT in the Republic of Korea, and Krasnoyarsk URL in the Russian Federation.

5.1.5. Spent fuel from non-power reactors

The amount of spent fuel from non-power reactors is much smaller than from nuclear power reactors. Fuel from non-power reactors, however, raises some specific challenges as it sometimes has higher enrichment than, and a different composition from, power reactor fuel. Although non-power reactors are often built in countries with a nuclear power programme, these reactors are also operated in countries without nuclear power plants where non-power reactor fuel is one of the most important factors in waste management. The IAEA works with Member States to develop a variety of nuclear education and training programmes, one of which is the Internet Reactor Laboratory. This is a cost effective virtual reactor that provides the possibility to use research reactors remotely, so the Member States without an existing research reactor can develop their nuclear infrastructure.

FIG. 10. Spent fuel disposal canisters, Posiva, Finland (courtesy of Posiva).

At present, most non-power reactor spent fuel is returned to the country of origin of the fuel, mainly the Russian Federation and the USA, and thus does not require disposal in the country where it has been used. A few countries, such as Australia, Belgium and Sweden, have decided to reprocess either part or all of their spent non-power reactor fuel. Some countries have to consider disposal of this spent fuel nationally and this might be a challenging task, especially if they do not have a nuclear power programme. IAEA Nuclear Energy Series NW-T 1.11 provides an overview of the available reprocessing and recycling services for non-power reactor spent fuel [31].

5.2. INTERMEDIATE LEVEL WASTE

ILW generally contains significant amounts of long lived radionuclides and therefore requires disposal at depths that provide isolation from the biosphere over the long term. ILW requires shielding during handling and storage. It has to be noted that the definition of ILW used in GSG-1 [3] is used throughout this publication, which means that the ILW covered in the publication includes all forms of ILW that require a greater degree of containment and isolation than near surface disposal can provide.

5.2.1. Processing

The processing of ILW either takes place at the facility where it is generated or at a purpose-built facility (which can also be a centralized facility). Processing consists of collection, segregation, decontamination, volume or size reduction and stabilization prior to packaging [12, 20, 21, 23]. Drying, evaporation, high pressure compaction, melting and cementing are common technologies applied in the treatment and conditioning of ILW [64]. Care needs to be taken during treatment to make sure that radioactivity concentrations will not increase beyond the capability of the treatment facilities or packaging to handle the resulting radiation levels and the extent of heat emission.

Depending on its intended storage or disposal destination, ILW is often treated and conditioned by incorporating it into a matrix (e.g. cement) within a suitable container to provide the required radiation shielding [65]. In some cases, where additional matrices are not required to ensure safety, conditioning is limited to packaging. In other cases, the waste object itself (such as a large vessel with internal contamination) forms the container, once suitably sealed.

Concrete containers with steel reinforcement, steel drums and steel boxes are commonly used for waste packaging. Their dimensions are selected to meet safety requirements and to be compatible with the dimensions of transport casks and disposal vaults. ILW containers can either be self-shielded or rely on external shielding to provide the necessary radiation protection. Both design concepts are used extensively.

5.2.2. Storage

After processing, storage of the product is often necessary if suitable disposal facilities are not available. Storage for periods of up to 100 years or longer can be considered as an option provided that the waste containers will remain intact and are not subject to degradation. Attention needs to be paid to the provision of adequate containment and shielding. Heat removal may also be required in some cases, although not to the same extent as HLW.

5.2.3. Disposal

The only licensed disposal facility for long lived ILW is the Waste Isolation Pilot Plant (WIPP), USA, where long lived, non-heat-generating waste from defence activities is disposed of in a geological repository built in salt beds. Elsewhere, ILW is held in storage until a disposal facility suitable for this material becomes available. Germany and Switzerland envisage that all LLW and ILW will be disposed of in one multipurpose, deep geological facility, obviating the need to separate waste containing short and long lived radionuclides before disposal. Germany plans to dispose of all types of radioactive waste in deep geological formations, with waste being classified either as heat generating or non-heat-generating, with separate repositories. In France, long lived ILW will be disposed of together with HLW in the planned Cigéo facility.

5.3. LOW LEVEL WASTE

Taken together, VLLW and LLW typically account for more than 95% of the volume but less than 2% of the radioactivity of all radioactive waste. LLW does not generally require significant shielding during handling and interim storage. The waste is suitable for disposal in engineered near surface facilities. However, some countries, such as Germany and Switzerland, are implementing policies that do not foresee separate disposal facilities for every radioactive waste class, and therefore their LLW might be disposed of at deeper facilities.

5.3.1. Processing

As with ILW, the treatment and conditioning of LLW either takes place at the facility where it is generated or at a purpose-built facility (which can be a centralized facility). The waste is segregated, treated, conditioned, packaged, monitored and stored, as appropriate, before being transferred to the disposal facility. Drying, incineration, evaporation, high pressure compaction, melting and cementing are common processes applied to the conditioning of LLW [66]. Concrete containers, steel drums and steel boxes are commonly used for waste packaging. Subject to meeting all relevant safety requirements, their dimensions are selected to fit the dimensions and shapes of disposal spaces and transport packages.

5.3.2. Storage

Options for the storage of LLW are broadly similar to those for ILW (see Section 5.2.2). Storage for periods of up to 100 years or longer can be considered as an option provided that the waste containers remain intact and are not subject to degradation. Such long term interim storage is implemented in the Netherlands.

5.3.3. Disposal

LLW, most of which has a half-life of less than 30 years, is disposed of in near surface repositories in many countries (see Annex 2 and Figs 11–13). These are trenches or concrete vaults into which containerized waste is placed. An engineered cover system is placed over the waste to limit water infiltration and surface erosion and to prevent intrusion by humans or burrowing animals. The facilities are subject to surveillance until the hazard associated with the waste has declined to acceptable levels (typically a few hundred years). While disposal of LLW in a near surface facility is a typical strategy for many, some countries (e.g. Canada, Finland, Germany, Hungary, the Netherlands, the Republic of Korea, Sweden and Switzerland) have chosen, or are considering, the option of disposing of LLW in repositories at depths between 50 m and 1000 m. These facilities do not require any long term surveillance. Several countries have both licensed and operated geological disposal facilities for LLW, either as purpose-built facilities (e.g. Bátaapáti repository in Hungary; Gyeongju facility in the Republic of Korea; Himdalen repository in Norway; and SFR, the Final Repository for Short-Lived Radioactive Waste, in Sweden) or converted facilities from former mines of various types (e.g. Richard in the Czech Republic; Asse II and Morsleben in Germany; and Băiţa Bihor in Romania).

FIG. 11. LLW silo in the disposal facility in Olkiluoto, Finland (courtesy of TVO).

FIG. 12. *VLLW and LLW disposal facility in El Cabril, Spain (courtesy of Enresa).*

FIG. 13. *Low and intermediate level radioactive waste disposal vault in Bátaapáti, Hungary (courtesy of PURAM).*

A small number of countries are considering LLW's co-location in geological facilities with ILW, HLW or spent fuel. Co-disposal can result in a simpler waste management system because fewer facilities need to be developed. However, co-location can also introduce design complexity to avoid interferences between the waste types (e.g. decomposition of LLW can result in the generation of complexing agents that reduce the safety of higher level waste), as well as significant increases in the volume of material requiring handling at geological depths.

5.4. VERY LOW LEVEL WASTE

VLLW often exists in large volumes and is mainly generated during the decommissioning of nuclear facilities or from the cleanup of contaminated sites. Typical VLLW includes concrete, soil and rubble. This class is currently recognized as a distinct classification by only a small number of States (e.g. France, Japan, Lithuania, Spain and Sweden). In most other country classification systems, it is included as part of the LLW stream.

5.4.1. Processing

VLLW is typically not subject to extensive processing, apart from its packaging, due to the very large quantities involved and the low content of radionuclides. In countries where the clearance concept is used, the volume of potential VLLW can be reduced by appropriate characterization to separate those components that can be released from regulatory control as cleared waste.

5.4.2. Storage

Generally, VLLW is stored at the site of its generation or in a centralized storage facility until it can be transported to a suitable disposal facility. During this stage, a simple shelter or temporary cover might be sufficient to provide protection from wind, rain, etc.

5.4.3. Disposal

In France, Slovakia and Spain, VLLW is disposed of in purpose-built disposal facilities in shallow trenches with engineered covers, often near the site of generation to avoid the transport of large volumes of material (see Fig. 14). Sweden and Lithuania developed an above ground design using a concrete slab. In other countries it is disposed of together with other waste types, such as LLW, or (in countries such as the UK) with non-nuclear hazardous waste.

5.5. URANIUM MINING AND MILLING WASTE, NORM WASTE

Uranium mine waste rock and mill tailings are normally managed and disposed of close to the uranium mine or the uranium mill. The waste rock and tailings are not packaged, but rather are contained in nearby locations with suitable barriers (stable mounds with an appropriate cover system) to minimize their radiological and non-radiological impact on the surrounding environment. In some countries, UMM tailings and in situ leaching waste are not classified as radioactive waste. Hence, these countries do not report the waste as radioactive waste, while others do. UMM can then be classified as long lived VLLW, or in some cases LLW. Uranium extraction by the in situ leaching method usually also generates smaller volumes of radioactive waste and different waste forms.

Radioactive residues are also generated from the oil and gas industries (e.g. scales and sludges), mining of other minerals and products (e.g. residues from extraction of thorium and rare earth elements), and the treatment and usage of drinking and process water.

If NORM is classified as radioactive waste, depending on the national waste management concept, this is usually considered to be VLLW or LLW. NORM waste is not specifically discussed in this publication, although some countries have reported NORM waste in their National Profiles.

5.6. DISUSED SEALED RADIOACTIVE SOURCES MANAGEMENT

Sealed radioactive sources are used widely in medicine, industry and agriculture and, because of this, they are found in almost all countries. For many countries, these are the only radioactive materials to be handled, and they require storage and eventually disposal. The life cycle of a sealed radioactive source is presented in Fig. 15.

On account of the special nature of DSRSs and their widespread use, specific international standards have been developed for their management, including the following:

- Code of Conduct on the Safety and Security of Radioactive Sources [8];
- Council Directive 2013/59/Euratom of 5 December 2013 laying down basic safety standards for protection against the dangers arising from exposure to ionising radiation, in particular Section 2 on control of radioactive sources [67];
- Council Regulation 1493/93/Euratom of 8 June 1993 on shipments of radioactive substances between Member States [68].

The Guidance on the Import and Export of Radioactive Sources was published in 2012 [9], and in 2018 supplementary Guidance on the Management of Disused Radioactive Sources [10] was developed and published, providing more details on the effective management of DSRSs.

Depending on the intended use, sealed radioactive sources include a wide variety of radionuclides and activity levels. The Code of Conduct [8] and IAEA Safety Standards Series No. GSR Part 3, Radiation Protection and Safety of Radiation Sources: International Basic Safety Standards [11], categorize radioactive sources according to their potential to cause serious health effects (see Table 5).

FIG. 14. VLLW disposal at the CIRES facility, France (courtesy of ANDRA).

TABLE 5. CATEGORIES OF SEALED RADIOACTIVE SOURCES [23]

Category	Risk in being close to an individual source	Examples of uses
1	Extremely dangerous to the person	Radioisotope thermoelectric generatorsIrradiators
2	Very dangerous to the person	Industrial gamma radiography sourcesHigh/ medium dose rate brachytherapy sources
3	Dangerous to the person	Fixed industrial gaugesWell logging gauges
4	Unlikely to be dangerous to the person	Bone densitometersLevel gauges
5	Most unlikely to be dangerous to the person	Permanent implant sourcesLightning conductors

At some point, sealed sources have to be replaced, usually because their activity level has declined below which the source is no longer suitable for its intended purpose. They are then considered to be 'spent' or 'disused' sources. DSRSs are either managed together with other waste in a category commensurate with their hazard (e.g. LLW, ILW or HLW) or separately, again in a manner commensurate with their hazard. An overview of the national strategies used in DSRS management can be found in Annex 5.

5.6.1. Storage and conditioning

States with nuclear power facilities are likely to have the capacity for long term storage or disposal of DSRSs together with other types of radioactive waste. For many small countries, however, storing or disposing of the sources safely and securely presents an ongoing challenge. The management practices for DSRSs are very similar to the management of LLW and ILW.

Sources with short half-lives (e.g. ^{192}Ir, half-life of 74 days) can be stored until the radioactivity in the source decays to low enough levels to allow release from regulatory control (i.e. clearance); while others (e.g. ^{226}Ra, which until recently was widely used) remain potentially hazardous for tens of thousands of years. Where disposal options are not available, long term storage facilities are required for many types of DSRs. Effective management involves repackaging the source, checking the condition of the source or source container regularly, and providing appropriate safety and security measures.

5.6.2. Return to supplier, reuse and recycle

In the case of disused sealed sources, the preferred option for managing them is recycling for further use. If this is not possible, the next preferred management option is the return of the source to its supplier [27]. As a result of the challenges associated with disposing of DSRSs safely, especially in countries with little or no radioactive waste management infrastructure, current good practice is to return the sources to the manufacturer for refurbishment, recycling or disposal. A number of countries insist upon this as a condition of the import and sale of sealed sources within their territory.

There are various recycling methods available, such as recovery of the sealed source. Recycling reduces the amount of radioactive material that needs to be produced; however, at the same time it needs to be taken into account that these actions have to be cost effective and technically feasible.

While the return of the DSRSs to the supplier is a widely used option, it is not always possible, as the original supplier may, for example, be unknown or no longer exist, or the transport means, regulatory framework or financial resources may not enable transportation of the sources.

5.6.3. Disposal

If no further use is foreseen and it cannot be otherwise removed from regulatory control, the only sustainable long term option is disposal. As such, disused sources for which no recycling or repatriation options exist have to be declared as radioactive waste and need to be managed as such, in compliance with relevant international legal instruments, safety standards and good practices.

For those disused sources that cannot be returned to a supplier or reused and that cannot be stored until they decay to clearance levels, disposal is the final step in their management. Some countries, particularly those with a nuclear power programme, may have the option to co-dispose their disused sources in a near surface or geological disposal facility. It will, however, need to be verified that the sources comply with the waste acceptance criteria set up for those facilities.

Where co-disposal is not possible, disposal in one or more boreholes may offer a solution. Disposal in boreholes offers a safe and secure disposal solution. The concept of borehole disposal of disused sources has been extensively studied and developed over the last two decades. In that time, it has evolved from a conceptual idea into a mature disposal solution. Today, projects on the borehole disposal of DSRSs are ongoing in Ghana and Malaysia and are being considered in several other countries.

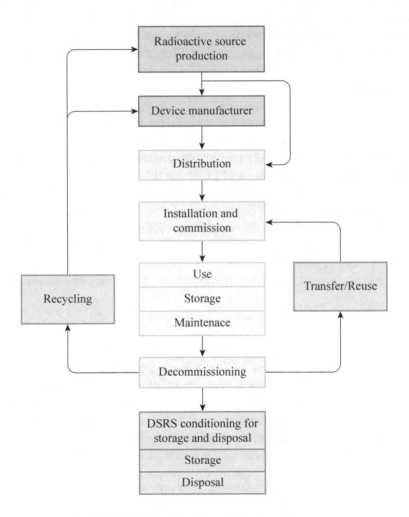

FIG. 15. The life cycle of a sealed radioactive source.

6. INVENTORIES

6.1. DATA SOURCES

The main sources of information used for inventories and forecasts are the National Profiles on the web site accompanying this publication. Additionally, publicly available Joint Convention National Reports were used. In the case of some EU members, their openly available National Reports on the implementation of the Euratom Waste Directive were used as well. The National Profiles, Joint Convention National Reports and EU National Reports cover almost 95% of all nuclear power plants in the world. This provides a good basis for making regional and global aggregations of waste volumes. Nuclear power plants generate significant quantities of spent fuel and radioactive waste, and by comparison, countries without nuclear power plants generally have much smaller amounts of radioactive waste (and of spent fuel if they operate research reactors).

The estimation of the global inventory depends on the availability of the inventory data for countries with nuclear power plants. All countries have some radioactive waste, which in many cases entirely comprises DSRSs. While the aggregated numbers have been rounded for presentation purposes, all waste quantities provided in the National Profiles, no matter how small, have been included in the aggregated global data quantities presented here. The United Nations country groupings were used for regional aggregation of the data.

6.1.1. National Profiles

A template for the National Profiles was generated to provide a structure for data collection that enabled aggregation of data at regional or global levels and facilitated data analysis. The template includes tables for amounts of spent fuel and radioactive waste, together with transformation matrices to enable the transfer of the volumes given in national classification systems to the classification in GSG-1 [3]. A total of 38 States contributed National Profiles, which are provided on the web site.

6.1.2. Joint Convention National Reports

Joint Convention National Reports are produced triennially and Contracting Parties to the Joint Convention are encouraged to publish their National Reports.[4] Data were used from this source for States that did not submit a National Profile.

The Joint Convention National Reports include listings of spent fuel and radioactive waste management facilities in the States. They also provide inventories of spent fuel and radioactive waste in the States based on their national waste classification systems. Forecasts of spent fuel and waste storage and disposal are, however, not provided. National Reports vary in the level of detail provided and in the measurement units used, requiring translation of the waste quantities presented according to national classification into equivalent GSG-1 waste classification.

6.1.3. Euratom Waste Directive National Reports

The Euratom Waste Directive requires submitting a triennial report to the European Commission on the implementation of the Directive. These reports give a comprehensive but concise high level overview of how a Member State complies with the Directive, with an emphasis on major changes and progress made since the previous report [69]. The inventory dates and timing of these reports are similar to the Joint Convention National Reports for several Member States.

---

[4] See https://www-ns.iaea.org/conventions/results-meetings.asp.

EU Member States are required to provide in their national programme an inventory of all spent fuel and radioactive waste and estimates for future quantities, including those from decommissioning, clearly indicating the location and amount.

6.2. DESCRIPTION OF DATA AGGREGATION

As noted previously, the precise definition of what constitutes radioactive waste and its classification into levels or categories varies widely among countries, which creates some inherent difficulties in aggregating inventory data regionally and globally. This section describes the approach taken to data collection and the model used to aggregate data from different countries into a common framework.

6.2.1. Conversion to IAEA waste classification

This publication uses the waste classification in GSG-1 [3] to present global values. This classification is based on the disposal route required to provide long term safety. However, some States intend to dispose of waste in facilities normally reserved for waste that presents a greater long term hazard (e.g. disposing of LLW in a geological repository). Furthermore, the boundaries between different classes are not defined by quantitative activity levels, but instead depend on the safety case for a specific facility. On that basis, waste of a particular class in one country might not have precisely the same level of activity as the same class of waste in another country — although the differences at the margins are typically not significant. Note that the total amount of all classes of waste remains the same and can be stated with a good degree of accuracy, and only its allocation between different classes or categories may be subject to variation.

To assist the conversion to the waste classification in GSG-1 [3], respondents were asked to include a conversion matrix as part of their National Profiles, indicating the proportions of waste from a national class corresponding to the appropriate waste classification in GSG-1. Estimates were made for States that did not provide a conversion matrix. A similar approach was used with data taken from the publicly available National Reports under the Joint Convention [2].

6.2.2. Constraints in determining global inventory

The data presented in the National Profiles has provided the basis for preparation of the global and regional aggregated data presented in this publication. However, this process involves some additional uncertainties. For example, the waste volumes can also be presented in different ways and the determination of 'as disposed' waste volumes requires assumptions to be made, which inevitably involves the use of approximations, concerning the waste processing and disposal strategies. This step tends to be particularly complex in the case of liquid waste.

The recognized gaps and uncertainties in the estimation of global inventory data include the following:

- Lack of data on some countries. This will result in an underestimate of total inventories.
- Uncertainties in the translation of data from national waste classification systems to waste classification in GSG-1 [3] for aggregation purposes. This will affect the distribution of waste volumes among the various waste classes (VLLW, LLW, ILW and HLW), but will not affect the overall total amount of waste.
- Differences in the way that various States report waste volumes (e.g. 'current as stored' volumes versus forecasted 'as disposed' volumes, use of actual physical volume of waste packages, versus the volume envelope it might occupy in a repository). This will affect the reported volumes of waste. On average, however, the overall effect on accuracy of the global inventories would not be significant due to offsetting increases and decreases as well as rounding of the aggregate numbers.
- Different reporting dates will affect the accuracy of a 'snapshot' for a given date. However, most of the reporting dates are within a year or two of the selected reference date for this publication

(31 December 2016). Given that in most cases the accumulated waste and spent fuel volumes do not grow very quickly, and given the very large residual inventories in countries with large programmes, the overall effect on the accuracy of the global inventories would not be significant.

- Different approaches to the clearance of radioactive waste.
- The inclusion of unprocessed liquid waste in the totals. In some cases, no distinction was made in country reports for unprocessed liquid waste versus solid waste. The potentially large volumes of liquid waste can distort the overall data if not accounted for separately. Therefore, where a country has distinguished between liquid and solid waste, either by direct statement or inference from a waste classification, liquid waste quantities are handled separately from solid waste quantities in all relevant tables of this report.

The project supporting the development of this publication did not include a quantitative analysis of the level of uncertainty in the presented information. Lessons learned in the collection and analysis of data for this publication will be incorporated into later phases, and modifications will be sought to improve accuracy and to minimize uncertainties in the aggregated data.

6.2.3. Conversion to disposal volumes

The template for the National Profiles requested waste volumes to be presented corresponding both to the current state of the waste and its anticipated volume for disposal. To help minimize the inconsistency in volumes, this publication uses 'as disposed' volumes where available, followed by 'as stored' when only this has been reported. Estimations are, however, necessary to calculate the 'as disposed' volume, taking into account the repository requirements and the conditioning and packaging plan.

For some countries, with a known or assumed conditioning and disposal route, it is possible to transform the storage volume to the disposal volume. For States without established plans for a repository and corresponding waste package geometries, several assumptions need to be made, for example concerning what further conditioning and packaging will be required for 'disposal ready' packages. In such situations, greater uncertainty may exist concerning the disposal volumes. This uncertainty might increase if there is a possibility that conditioned waste packages are eventually placed in larger containers or overpacks for disposal.

The Status and Trends project includes an initiative by the IAEA, the OECD/NEA and the EC to harmonize the spent fuel and radioactive waste inventory data reported to the different agencies for various purposes to reduce the reporting burden on Member States and to ensure the consistency of data reported. It was agreed that the volumes of conditioned waste ready for disposal need to be used. This is also recommended by the European Nuclear Safety Regulators Group (ENSREG) [69] and OECD/NEA Expert Group on Waste Inventorying and Reporting Methodology (EGIRM), which has been developing a methodology to provide a scheme for presenting spent nuclear fuel and all types of radioactive waste and corresponding management strategies that could be included in inventories worldwide [70]. The following main entities were suggested by the EGIRM:

- Waste form (WF): Radioactive waste or spent fuel, in the form in which it will be disposed of, including any stabilizers and excluding the waste container;
- Waste package (WP): The ensemble of one or more WF together with its waste container, which has to be suitable for handling and storage, and may allow transport and disposal if waste acceptance criteria are addressed;
- Disposal module (DM): The ensemble of one or more waste packages together with their disposal containers (overpackage) and optional buffer; may be suitable for handling, transport and storage and needs to be suitable for disposal without further conditioning.

It is noted that there is no universal agreement on the definition of 'disposal volume'. For any given country, the definition and/or calculation method is usually embedded in regulations, national policy or a facility licence. Some countries consider the DM to be part of the disposal facility, not a property of the waste. In this case, the disposal volume corresponds to the WP volume. Other countries consider the DM to be the final WP and interpret the disposal volume as the volume of the DM. In both cases, some countries interpret the volume as the actual physical volume for the package or module, while others base it on an envelope volume or packing factor to take account of inefficiencies in package stacking in the disposal facility. Figure 16 illustrates the different terms used.

6.3. CURRENT INVENTORIES OF SPENT FUEL

The spent fuel inventories provided in this publication do not distinguish between fuel that is considered to be a waste in the responding State and fuel that is considered to be an asset (i.e. intended to be reprocessed). The global totals include all countries where information is available. The data include spent fuel from nuclear power plants, demonstration and research reactors and other kinds of reactors (e.g. isotope production). The amount of spent fuel is presented in tonnes of heavy metal (t HM) and describes the mass of heavy metals (e.g. plutonium, thorium, uranium and minor actinides) contained in the spent fuel.

It is noted that spent fuel that has been sent for reprocessing but has not yet been reprocessed is included in the amount of spent fuel currently in storage in the country to which it has been sent. Additional data on historical amounts of spent fuel that have been reprocessed have been extracted from other sources, such as the annual reports from commercial reprocessing facilities. Spent fuel that has been reprocessed is no longer in the form of fuel, but has been separated into various types of waste and recyclable components. The unit of measure for waste from the reprocessing of spent fuel is the cubic metre, and is included as part of the LLW, ILW and HLW as appropriate. Fuel that has been reprocessed is included separately in the tables under 'sent for reprocessing' to give a total of all the spent fuel that has been produced since the beginning of the nuclear power age.

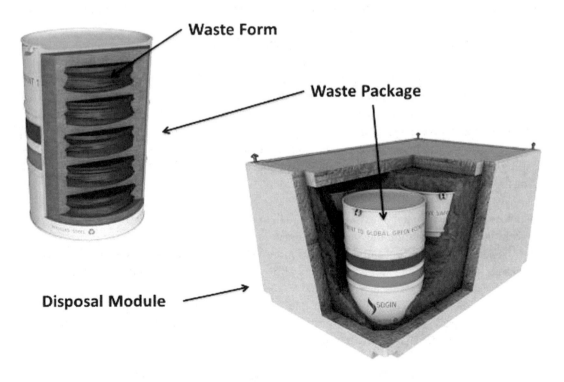

FIG.16. Waste form, waste package and waste disposal module [70].

6.3.1. Nuclear power plant spent fuel

Between the start of nuclear-power-based electricity production in 1954 and the end of 2016, a total of about 390 000 t HM of spent fuel was discharged from all nuclear power plants worldwide (see Table 6). The aggregation in Table 6, as well as those in subsequent tables and charts, gives a global summary of inventories. It is also divided into subtotals by UN country geographic regions as well as giving a global total. It shows subsets of the global total for the Member States of the Joint Convention, EU and OECD/NEA. These subgroupings allow the reader to see the aggregated amounts for various Member State groupings. Details for individual countries can be found in the Country Profiles located on the web site accompanying this publication.

About one third of all spent fuel discharged from nuclear power plants (127 000 t HM) has been reprocessed. The remaining two thirds are stored, pending processing or disposal. Most spent fuel is held at nuclear power plant sites in wet storage in the reactor pools. Fuel inside the reactor core is not included in the inventory, since it is not considered to be spent until it has been discharged from the core.

After initial storage for cooling for at least a few years in the reactor pool, some spent fuel has been transferred to dry storage or to centralized wet storage facilities. The total amount of spent fuel in storage was about 263 000 t HM as of the end of 2016. Figure 17 shows the share of spent fuel stored either in dry or wet storage.

6.3.2. Spent fuel from research and other reactors

A number of States operate non-power reactors, such as research, isotope production, experimental, prototype or propulsion reactors. The national inventories of spent fuel from these reactors are summarized in Table 7. It is noteworthy that these amounts are less than 1% of the amount of spent fuel that originates

TABLE 6. REPORTED SPENT FUEL DISCHARGED FROM NUCLEAR POWER PLANTS, AS OF 31 DECEMBER 2016

Region	Wet storage (t HM)	Dry storage (t HM)	Reprocessed (t HM)	Total (t HM)
Africa	950	50	n.a.[a]	1000
Americas	83 500	52 500	600	136 500
Asia	35 500	6500	8500	51 000
Europe	63 500	20 500	117 500	201 500
Oceania	1	n.a.[a]	1	1
Global total	183 500	80 000	127 000	390 000
Joint Convention Contracting Parties	180 000	80 000	127 000	386 000
EU Member States	42 000	11 000	113 000	166 000
OECD/NEA members	163 000	72 000	126 000	362 000

Source: National Profiles and Joint Convention National Reports.
Note: Possible differences in totals are due to rounding.
[a]n.a. not applicable (or none reported).

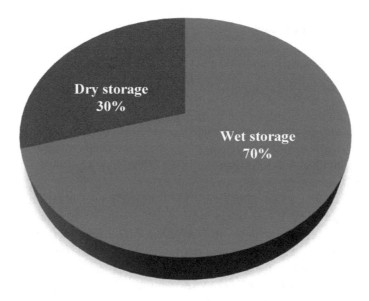

FIG. 17. Nuclear power plant spent fuel storage by type.

from nuclear power plants. It is noted that the fuel quantities from non-nuclear power plant reactors are generally not publicly reported to the same level of detail as for nuclear power plants. However, a typical research reactor has a core capacity in the order of a few kilograms of uranium fuel, whereas a commercial nuclear power plant might have a core of 100 tonnes or more. Isotope production reactors may have a capacity of a few tonnes.

There can also be a difference in the enrichment of the fuel in ^{235}U. Many research reactors and isotope production reactors were originally designed to operate using high enriched uranium (HEU) fuels, whereas power reactors utilize low enriched uranium fuels (roughly 0.7–5%), leading to a potential difference in the amount of uranium that is discharged in the spent fuel. This also depends on the amount of 'burnup' of the ^{235}U. In many cases, the research reactors have been converted to operate with low enriched uranium. For example, the policy of the USA has been the minimization, and ultimately elimination, of HEU in civilian research reactors worldwide since 1978 [71].

The majority of the spent fuel in storage from non-power reactors is in North America. This is because spent fuel from prototype power reactors is considered under the research reactor category in Canada and the USA. Most spent fuel from research and other reactors in many countries has been returned to suppliers for reprocessing or disposal (usually the USA or the Russian Federation), and in these cases that spent fuel will become part of the inventory of the receiving country.

6.3.3. Planned management of spent fuel

The planned management of spent fuel is summarized in Fig. 18, based on the fraction of the total tonnes of heavy metal of spent fuel currently in storage. It is possible that a State might change its strategy (e.g. owing to a change in national policy), in which case they might have a combination of reprocessing waste and spent fuel for disposal. In some States, different routes are planned for the spent fuel from different types of reactor, again resulting in a combination of reprocessing waste and spent fuel for disposal.

Waste from reprocessing is included in the inventory of radioactive waste in the country where it is currently stored. In most cases, when reprocessing takes place in a different country, the resulting HLW is eventually returned to the country from where the fuel originated.

TABLE 7. REPORTED SPENT FUEL STORED FROM RESEARCH AND OTHER REACTORS, AS OF 31 DECEMBER 2016

Region	Wet storage (t HM)	Dry storage (t HM)	Total (t HM)
Africa	0.1	0.2	0.3
Americas	39	2920	2959
Asia	109	n.a.[a]	109
Europe	1014	47	1061
Oceania	1	n.a.[a]	1
Global total	1163	2 967	4130
Joint Convention Contracting Parties	1163	2 967	4130
EU Member States	960	24	984
OECD/NEA members	1162	2967	4129

Source: National Profiles and Joint Convention National Reports.
Note: Possible differences in totals are due to rounding.
[a]n.a. not applicable (or none reported).

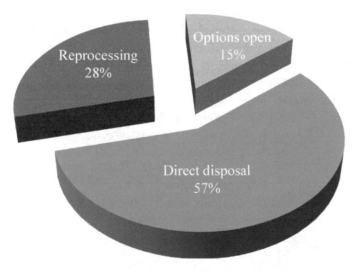

FIG. 18. Summary of existing spent fuel by planned management route.

6.4. CURRENT INVENTORIES OF RADIOACTIVE WASTE

Most of the radioactivity present in radioactive waste (up to 95% of the total) is present in HLW (including spent fuel, when declared as waste). In terms of volume, the situation is reversed and more than 95% of the total volume of waste comprises LLW or VLLW. The hazard presented by any toxic agent is a complex combination of the quantity, the particular chemical components and their respective concentrations in the waste (in this case mainly the radionuclides), the physical and chemical form of the waste, the radioactivity level and the exposure scenarios. Generally, chemical and physical forms of waste that are mobile in the environment are more hazardous. Limiting its mobility is therefore an important reason for conditioning waste prior to disposal, as well as for selecting suitable geology when siting a disposal facility.

Solid waste and liquid waste are described separately. This differentiation is important because of the significant volume of liquid waste and the large volume reduction achievable from processes such as evaporation, filtration, vitrification and others, depending on the chemical composition and amount of water. Typically, liquid waste is processed for solidification soon after it has been generated, rather than placing it in storage. In States that follow this approach, only a small part of the national radioactive waste inventory exists in liquid form. In some countries, a past practice was to store some waste in liquid form with the intention of processing and converting it to solid form at a later stage, and as a result, the waste still exists in this form at many sites. Section 6.4.2 presents the liquid radioactive waste inventories as indicated in the National Profiles submitted.

Most radioactive waste is either in storage awaiting the development of a suitable disposal facility, awaiting further treatment pending disposal in a licensed facility, or has already been disposed of. In general, only solid waste is placed into disposal facilities, although past practices in some countries included direct injection of liquid waste into underground formations for disposal. This strategy is still practised in the Russian Federation (see Section 6.4.2) and is also practised in non-nuclear industries in many countries, such as for the disposal of waste from oil and gas extraction, which may contain important concentrations of naturally occurring radionuclides.

Disposal is defined as intentional emplacement in a facility without the intent to retrieve. Although some States require the possibility of retrieving the disposed waste for some period of time after disposal, this is still considered to be disposal for the purposes of this publication.

6.4.1. Solid radioactive waste

Solid waste includes inherently solid materials, such as metals, plastics and other dry materials, as well as solidified liquids. In the case of unprocessed waste, and in some States, 'solid waste' can also include small amounts of liquids or 'wet solids' (such as filter cake or dewatered ion exchange resins). Figure 18, based on the National Profiles, shows the global totals of different types of solid radioactive waste in storage and disposal, as of 31 December 2016. The data shown in Fig. 19 represent 'as disposed' volumes, based on the conversion matrices provided by respondents (see Section 6.2.3 for a discussion of the uncertainties inherent in such an approach).

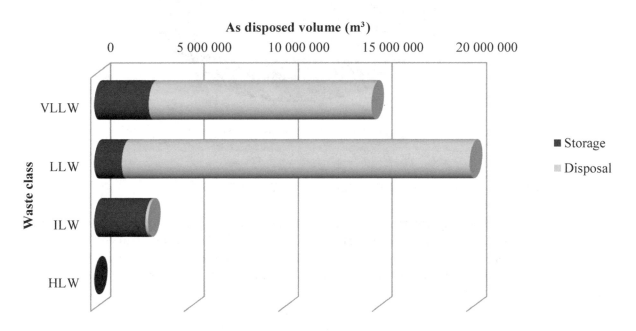

FIG. 19. Summary of reported global solid radioactive waste inventories (m³). HLW storage volume is 29 000 m³.

It is evident that the majority of the volume of waste consists of VLLW and LLW. VLLW has a smaller volume than LLW because often this category of waste was created more recently in a country's classification system and was not retroactively applied to radioactive waste that was already disposed of. As noted above, most of the radioactivity is contained in the much smaller volumes of ILW and HLW. For VLLW and LLW, the majority of the waste generated has already been disposed of. For ILW and HLW, however, the majority of the waste so far generated is currently in storage awaiting the development of appropriate disposal facilities.

Table 8 summarizes the volume of solid radioactive waste in storage and Table 9 summarizes the volume of solid radioactive waste in disposal, as of 31 December 2016. The values for the waste classes are as defined by the individual countries and the totals have been rounded to the nearest 1000 m^3.

It can be seen that 85% of the already generated VLLW and LLW has been disposed of, and actually most of the generated VLLW and LLW has disposal options available. For ILW, the fraction already disposed of is less than 10%, and for HLW there is no disposal option available. Since the 2013 edition of this publication, the fraction of disposed VLLW and ILW has increased, and the fraction of disposed LLW has decreased.

6.4.2. Liquid radioactive waste

While most States process liquid radioactive waste into solid form within a short time of it being generated, a few — notably, the USA and the Russian Federation — have large volumes of liquid waste in long term storage. Much of this waste results from defence activities and is only now being dealt with through the design, construction and licensing of liquid waste treatment facilities.

Table 10 summarizes the quantities of liquid radioactive waste declared by various countries. Note that these are the current 'as stored' volumes. It generally does not include liquids held in short term storage awaiting processing. It is difficult to estimate the final 'as disposed' volumes, since this will largely depend on the eventually selected processing and conditioning methods. In most cases, this will result in a very large reduction in the final volume for disposal, assuming that evaporative or filtering type processes will be used to separate or concentrate the radioactive elements from the bulk liquid. In other cases, the liquid waste might be conditioned in situ (e.g. by cementation), which can result in a volume increase. In addition, the classification of the final WF (VLLW, LLW, ILW or HLW) will also depend on the length of time (i.e. radioactive decay time) that the waste has been in storage, the efficiency of the treatment process and the degree of volume reduction achieved.

TABLE 8. REPORTED SOLID RADIOACTIVE WASTE IN STORAGE, AS OF 31 DECEMBER 2016

Region	VLLW (m^3)	LLW (m^3)	ILW (m^3)	HLW (m^3)
Africa	14 000	25 000	1000	0
Americas	2 309 000	303 000	85 000	6000
Asia	350 000	249 000	69 000	6000
Europe	245 000	890 000	2 583 000	22 000
Oceania	0	4000	0	0
Global total	2 918 000	1 471 000	2 740 000	29 000
Joint Convention Contracting Parties	2 916 000	1 471 000	2 740 000	29 000
EU Member States	245 000	392 000	321 000	6000
OECD/NEA members	2 868 000	1 100 000	1 430 000	28 000

Source: National Profiles and Joint Convention National Reports.
Note: Possible differences in totals are due to rounding.

TABLE 9. REPORTED SOLID RADIOACTIVE WASTE IN DISPOSAL, AS OF 31 DECEMBER 2016

Region	VLLW (m^3)	LLW (m^3)	ILW (m^3)	HLW (m^3)
Africa	0	14 000	0	0
Americas	11 041 000	15 392 000	91 000	0
Asia	800	67 000	0	0
Europe	369 000	3 002 000	43 000	0
Oceania	432 000	24 000	0	0
Global total	11 842 000	18 499 000	133 000	0
Joint Convention Contracting Parties	11 842 000	18 405 000	133 000	0
EU Member States	369 000	2 101 000	12 000	0
OECD/NEA members	11 841 000	17 581 000	103 000	0

Source: National Profiles and Joint Convention National Reports.
Note: Possible differences in totals are due to rounding.

The volume of liquids to be processed is very high, in the range of 60 million m^3, including 6.7 million m^3 of high level liquids, located mainly in the Russian Federation and in the USA.

The volume of liquid waste that has been disposed of by injection into deep wells, based on data provided in the National Profiles of the Russian Federation and the USA, is around 62 million m^3. In the following sections this liquid waste will be considered as a specific waste management issue that needs a specific management approach.

6.5. FUTURE FORECASTS

In order to plan adequately for the long term management of radioactive waste, it is necessary to forecast the waste quantities expected in the future. This is a task that needs careful attention, especially for States with numerous and diverse activities that result in the creation of radioactive waste and with many different organizations involved in producing and managing it. For many States, radioactive waste generation is closely related to electricity production from nuclear power plants, and thus future forecasts are closely related to predictions of the future use of nuclear power.

These countries, which were nuclear pioneers and were involved in the initial development of nuclear power, may also need to deal with significant quantities of waste associated with the decommissioning and remediation of these early facilities and sites. Usually the amounts of waste arisings are much higher than similar waste arisings from the present day nuclear power generation industry. Higher waste arisings have been noted on several occasions, including in China, France, the Russian Federation, the UK and the USA.

It is nevertheless important to make predictions regarding future waste arisings and to update them at regular intervals. This is important for the planning of facilities needed for storage, treatment and disposal and for establishing adequate funding for future waste management. It is recognized that precise numbers are not required to establish a reasonable basis for predicting future needs, as long as the inherent uncertainty in the quantity of future waste arisings is acknowledged. In due course, more precise data on waste volumes and radioactivity levels will be needed at the time of licensing of such facilities.

TABLE 10. REPORTED LIQUID RADIOACTIVE WASTE IN STORAGE, AS OF
31 DECEMBER 2016

Country	VLLW (m^3)	LLW (m^3)	ILW (m^3)	HLW (m^3)	Total (m^3)
Argentina	0	30	0	0	30
Armenia	0	0	2460	0	250
Australia	0	0	15	0	15
Bulgaria	0	5655	0	0	7000
Hungary	1581	3953	2372	0	8000
Indonesia	0	309	0	0	300
Lithuania	0	60 500	0	0	61 000
Russian Federation	0	322 000 000	98 000 000	6 330 000	62 006 000
Slovakia	0	3360	0	0	3000
USA	0	200	0	340 000	340 000

Source: National Profiles and Joint Convention National Reports.
Note: Possible differences in totals are due to rounding.

Defining the planning assumptions also requires proper consideration and attention. A major aspect to consider is the timeframe for the forecast: the longer the forecast, the less accurate it will be. The amounts and composition of waste from different practices will vary according to the following:

— How facilities operate;
— Industrial and technical processes that generate the waste;
— Policy and regulations governing the industry, technology and waste management;
— The economics of different areas of the waste management cycle and waste management philosophies.

In developing forecasts of future waste quantities, the following considerations need to be addressed:

(a) Planning scenarios for future generation of electricity from nuclear energy (e.g. high, low, and best estimates and constraints);
(b) Future operating strategies for the facility, such as the merits of waste minimization versus cost minimization;
(c) Timing of activities that impact on waste arisings, such as short term versus delayed strategy for remediation actions and solutions;
(d) Practical tools for creating forecasts (e.g. historical data, data from similar countries and facilities, and engineering estimates);
(e) Modelling and process mapping (e.g. understanding the route through different waste management systems and the impacts of this on forecasts);
(f) Accuracy requirements (e.g. significance of impact on the waste owner or WMO if estimates are significantly too high or low).

In general, a 'bottom up' forecasting approach provides the greatest accuracy and facilitates the highest degree of flexibility. Using this approach, individual waste streams or categories are estimated at a facility level and then aggregated with other waste streams to produce an overall estimate. Initial estimates for planning purposes can also be derived by extrapolating past history, taking into account the number of facilities operated and their lifetimes; this latter approach is often simpler and more pragmatic. Estimates of future waste arisings need to include all activities and life cycle phases of a facility that result in the production of radioactive waste, such as operation, maintenance, refurbishment and decommissioning.

Compared to LLW and ILW, spent fuel arisings are somewhat easier to forecast. The amount of spent fuel produced is broadly proportional to the amount of energy extracted from it. With knowledge of the number and type of reactors and their historical fuel burnup levels, a reasonable forecast can be made of future spent fuel arisings for the remaining lifetime of a reactor.

In some countries, the forecasting of waste volumes has been performed over many years and is regularly published. Other States have only recently begun to undertake rigorous forecasting activities. Consequently, not all States have reported forecasts in their National Profiles. It is noted that the Joint Convention [2] does not require the reporting of forecasts, only values of presently stored and disposed waste. However, the Euratom Waste Directive [4] requires that forecasts of the future generation of radioactive waste be reported.

Annex 6 provides a summary of the current and future amounts of spent fuel in storage and disposed of for 17 States, as provided in the National Profiles. Future updates of this publication will build on these initial estimates to provide a more comprehensive picture. Annex 7 provides forecasts for spent fuel and for different classes of radioactive waste (converted from national classification schemes to GSG-1 [3], where necessary), together with summary information on the main assumptions used for the predictions.

7. ANALYSIS AND ACHIEVEMENTS

The first publication in this Status and Trends series [1] was more focused on collecting the information about the spent fuel and radioactive waste management practices and inventories, while this second publication provides an additional analysis of the global trends. The discussion also includes significant achievements since the previous publication; discussion and analysis of current and emerging issues; and reports on progress in addressing previously identified issues. This section focuses on a discussion of the achievements and general issues related to spent fuel (whether it has been declared to be a waste or not) and to the various radioactive waste categories.

7.1. COMPARISON OF INVENTORIES IN 2013 AND 2016

Table 11 summarizes the changes in inventories between the first and second Status and Trends report. It can be expected that spent fuel and radioactive waste inventories worldwide will increase over the next years and decades. Spent fuel and radioactive waste from regular operation of existing and future nuclear installations will probably increase at about the same rate as it does today, while the increasing activity in decommissioning activities and remediation of legacy sites will generate a much higher amount of radioactive waste, in particular for the LLW and VLLW classes.

Between the end of 2013 and the end of 2016, about 20 000 t HM was discharged from reactors, totalling an accumulated amount of 390 000 t HM. Since 2013, the total amount of spent fuel discharged has increased by about 6%. Approximately one third of all spent fuel discharged from nuclear power plants (127 000 t HM) has been reprocessed to date. This is similar to the ratio in 2013. However, it is expected to decrease in the short term as the UK reprocessing facilities cease operation. Once new reprocessing facilities become available in China, Japan and the Russian Federation, the ratio of reprocessed fuel may increase again. The remaining two thirds of the spent fuel is stored pending processing or disposal.

TABLE 11. SUMMARY OF CHANGES IN REPORTED SOLID RADIOACTIVE WASTE INVENTORY DATA

Solid waste	2013 data			2016 data		
	Storage	Disposal	Total	Storage	Disposal	Total
VLLW (m^3)	2 356 000	7 906 000	10 262 000	2 918 000	11 842 000	14 760 000
LLW (m^3)	3 479 000	20 451 000	23 930 000	1 471 000	18 499 000	19 970 000
ILW (m^3)	460 000	107 000	567 000	2 739 000	133 000	2 872 000
HLW (m^3)	22 000	0	22 000	29 000	0	29 000
Total	6 317 000	28 464 000	34 781 000	7 157 000	30 474 000	37 631 000
NPP spent fuel (t HM)	Storage	Reprocessed	Total	Storage	Reprocessed	Total
	250 000	120 000	370 000	263 000	127 000	390 000

When compared with data provided at the end of 2013, the following can be seen:

- The total amount of solid waste (VLLW, LLW, ILW and HLW) in storage or disposal has increased by about 7%. This is comparable to the percentage increase in spent fuel discharges over the same period.
- The total amount of solid waste disposed of increased by about 7% (2 000 000 m^3), mostly in the VLLW category.
- Note that the apparent decrease in the amount of LLW disposed of, as reported in the 2016 data, is due to a recalculation of how some national waste classification systems map into the IAEA GSG-1 [3] classifications used in the aggregations and show up as a corresponding increase in the VLLW and ILW (mainly in the USA and Ukraine). The recalculation has also resulted in an artificially high increase in the amount of ILW in storage. This points out some of the inherent difficulties in aggregating data over multiple countries with diverse domestic classification systems. The key point is that each country has a classification system suitable for its own waste management needs, and the classifications used in country A need not be compared to those in country B.

The development of a comprehensive global inventory of liquid radioactive waste is still ongoing. Consequently, no comparison can be made to the previous publication, except to say that more countries are including liquid radioactive waste in their official inventories. Major effort is needed to increase awareness, at both international and national level, of the need to develop disposal facilities in a timely manner to prevent transferring such liability to future generations. The bulk of the liquid waste is located in the Russian Federation and the USA, and in both cases is mainly the result of past practices and activities. Efforts are under way in both countries to address this legacy waste.

7.2. MANAGEMENT OF SPENT FUEL AND HIGH LEVEL WASTE

Information about the implementation of the back end of the nuclear fuel cycle is presented in Section 5. Several countries have made decisions on the policy to recycle or to dispose of spent fuel; however, the options are kept open or the topic is still under discussion in many countries. There are different factors that influence Member States in choosing an open or closed cycle.

A theoretical study by OECD/NEA [42] looked at the economics of the back end of the nuclear fuel cycle, comparing higher level idealized systems of open and closed cycles. The main factors that influenced Member States in choosing their spent fuel management strategy could be divided into the following groups:

- Political/social;
- Strategic;
- Economic;
- Environmental impact;
- Non-proliferation/security considerations.

The spent fuel management policy is defined at the national level, even if nuclear operators are privately owned companies. The main factor leading to the choice of a spent fuel management policy in the frame of a long term vision is usually the nuclear power policy and strategy. Several countries have acknowledged that policy may change based on developments in technology, economics, public perception and environmental considerations. Flexibility in a spent fuel management strategy and/or nuclear waste management strategy may be an additional desired factor in the decision making processes, especially in the case of newcomer States to nuclear power.

In general, countries with very large nuclear programmes, such as France, the USA, Japan, China, the Russian Federation and the UK, have implemented their spent fuel management strategies with a long term view based on country-specific considerations. The USA and UK are following an open cycle policy, whereas others have made the choice of the closed cycle with the long term view of the implementation of Generation IV reactors. Delays in such implementation could occur and may raise some concerns related to the management of the recovered materials if not recycled in the Generation III reactor fleet. Countries choosing the open cycle usually have no immediate interest in or purpose for using the uranium and plutonium recovered from the reprocessing of the spent fuel.

The approaches of small to medium size nuclear power countries are diverse. Most of them have opted for the open cycle route, considering both non-proliferation and economic aspects. This is the case, for instance, for Sweden and Finland. Some others, like the Netherlands, have chosen the closed cycle route with full recycling of the valuable materials, the drivers being the cost certainty and the environmental impact of storing/disposing of HLW instead of spent fuel. This also provides additional flexibility towards possible shared disposal solution in the future. Looking at possible future flexibility, other countries, such as the Czech Republic, Hungary, Slovenia and South Africa, have chosen the open cycle as a reference but are examining alternatives.

Some countries, including Argentina, Belgium, Brazil and Ukraine, have chosen to implement storage of spent fuel while maintaining the flexibility to decide in the future on an open cycle or closed cycle policy. However, it needs to be kept in mind that storage is just an intermediate step of the spent fuel management and cannot be considered as an end point.

Although the rate at which spent fuel is currently reprocessed is more or less constant, the trend lately has been that fewer States send their spent fuel for reprocessing overseas, and that the amount of spent fuel in long term storage is increasing. Reprocessing capacities have been lowering due to the fact that there are a few new plants in construction and several reprocessing plants are closed or nearing the end of their lifetimes. The development of the nuclear programmes in China, India and the Russian Federation could change this trend.

Long storage periods introduce several challenges in terms of technical aspects (e.g. changes of technologies, ageing management of facilities, etc.), licensing (changes of regulation), organization (changes in the nuclear industry) and funding (accuracy of costs in the long term and availability of funds), and the challenge is thus to ensure the long term safety and integrity of the storage facilities and the spent fuel/HLW for many decades to come. However, such challenges are properly managed and countries have decades of experience with spent fuel storage, both in wet and dry storage facilities. There are also active programmes in place to monitor the condition of the spent fuel and its storage environment to ensure it can be safely stored for the required length of time. This already motivates the improvement of existing storage facilities, for example in Finland. Requirements and issues related to spent fuel storage are discussed in further detail in IAEA Nuclear Energy Series No. NF-T-3.3 [29], IAEA-TECDOC-944 [72] and IAEA-TECDOC-1862 [73].

As the requirements for storage capacity increase, new storage is built outside of the reactor buildings. Due to the need for longer storage, there has been successful implementation of storage facilities, which are planned, built and operated either in the vicinity of the reactor building or as a centralized facility in the country. Most of these are facilities for dry storage, but some pool facilities are also in operation. Depending on the strategy for spent fuel management, AFR facilities using wet or dry storage technologies have been licensed and built. There are AFR on-site facilities, for example, in Hungary, Belgium, Canada, Spain and the USA, and centralized off-site AFR facilities in Switzerland, Germany, Sweden and the Netherlands. Currently, around 80% of the AFR facilities are based on dry technologies, and this is mainly due to their modular and passive nature. In order to be economic, a wet storage pool generally needs to be large and hence has a fixed capacity. Dry storage facilities, especially of the cask type, can be built to any size scale and can be expanded incrementally over time. This means that not all the cost is required up front as it is in a fixed capacity wet pool.

Some recent progress and achievements in the development of deep geological disposal facilities include the following:

- In Finland, the construction licence for the Onkalo site was granted in 2015. This is a first of a kind in the frame of a DGR. The construction of underground access tunnels at Onkalo started in December 2016 and the construction of the encapsulation plant started in 2019. The operation licence application is planned to be submitted in 2021.
- The licensing process is ongoing in Sweden (application made in 2011). After six years of review by the Swedish regulator, Sweden held its main hearings in autumn 2017 on the licence application for a spent nuclear fuel repository at Forsmark and an encapsulation plant at Oskarshamn. The Land and Environment Court held hearings over a combined total of five weeks, beginning on 5 September 2017, in Stockholm, Oskarshamn and Östhammar. In January 2018 the Court issued its findings on the environmental licensing process to the government. The major aspects of the application (e.g. issues surrounding the Forsmark site, including the rock, the buffer and the environmental impact assessment), as well as the encapsulation plant in Oskarshamn and increased capacity in the storage facility at CLAB, were approved. Additional information was requested by the Court regarding the properties of the canister and long term safety.
- In France, an Act in July 2016 specified the procedures for the creation of a reversible deep geological disposal facility. A review of the safety options report was made in 2016 and the regulator issued its conclusions in 2018. The commissioning of the facility is expected to start in 2025 with inactive operations transitioning to the start of the active operations in 2030.
- In Canada the Nuclear WMO (NWMO) is responsible for designing and implementing Canada's plan for the safe, long term management of used nuclear fuel. In June 2007, the Government of Canada selected adaptive phased management as Canada's plan for the long term management of used nuclear fuel, which requires spent fuel to be contained and isolated in a DGR. A total of 21 potential host communities started the voluntary site selection process in Canada, and five remain in the process after initial screening studies. Borehole drilling started at a potential repository siting area in Canada in late 2017. It is planned to continue borehole drilling and expand field studies to inform the assessment of geoscientific, engineering, environmental and safety factors. NWMO expects to select a preferred single site for the repository by about 2023.
- Switzerland is in stage 3 of a three stage siting process. Three potential siting locations have been identified (Jura Ost, Nördlich Lägern and Zürich Nordost) and further site characterization work is ongoing, including borehole exploration. A preferred site is expected to be identified by the early 2020s, with a general licence application submitted around 2024.

- The German siting process for the geological disposal facility in Gorleben was relaunched after suspending the previous studies. The repository site selection procedure is divided into three phases: the identification of potential host rock and selection of site regions for surface exploration; the exploration of the selected site regions and selection of sites for underground exploration; and the underground exploration and determination of the site by the German Bundestag and Bundesrat.

- Consultation on a new DGR multistage siting process in the UK commenced in January 2018 under government direction, with broad public consultation and dialogue, based on finding a volunteer host community and suitable geology. At the beginning of 2020 the approach to DGR site evaluation in England and Wales, following a comprehensive and open national consultation, was published.

- A DGR programme has been established in the Czech Republic. It is envisaged that the Czech repository will be constructed in a suitable granite rock mass approximately 500 to 1000 m below ground level. It is planned that a site will be selected by 2030 and construction work will commence in about 2035. Preliminary geological surveys were started at seven potential sites in 2015. Four candidate sites were selected in 2020 for further study, to be reduced to two by 2020. The underground research laboratory in Rožná mine in Bukov municipality has been in construction since 2017.

- In the Russian Federation efforts were under way on the establishment of an underground research laboratory at a potential DGR site (Zheleznogorsk). In 2016 expert evaluation was performed for the siting and construction of a site specific underground research laboratory in the Nizhnekansk rock massif.

- The Nuclear WMO of Japan (NUMO) published a geological screening map for a DGR in 2017, identifying potentially suitable areas as well as exclusion areas. A three stage, consent based siting process is being established. It is currently in stage 1 (literature survey). The second stage is preliminary investigation, and the final stage is detailed investigation. The overall process is expected to take about 20 years before a preferred site is identified.

- The Chinese policy on HLW disposal is that the spent fuel from LWRs is first to be reprocessed, and the waste arisings from this process are then to be vitrified and finally geologically disposed. The Chinese strategy for HLW disposal is characterized by three typical stages: laboratory studies and site selection for the HLW repository (2006–2020); underground in situ testing (2021–2040); and repository construction (2041–2050). China has started construction of an underground research laboratory in a crystalline rock formation in 2021 (Beishan area in Gansu Province, China).

The site selection approaches taken by the Member States can be different — some are taking a technical siting approach, while others are taking a voluntary approach or a combination of technical and voluntary. In all cases, the importance of public acceptance has been recognized, and all of the programmes include extensive public consultations at various points in the process. Thus, the DGR schedules have been revised to take into account realistic timescales for the siting activities, including seeking public acceptance, technical implementation and regulatory activities due to challenges caused by lack of experience of licensing such facilities, etc. Several countries have had to restart their site selection process for a DGR, having not been successful in gaining public support. Proposed timelines for the operation of DGRs vary from country to country, but to date the estimated soonest and latest years for opening DGRs are 2024 and 2160, respectively.

It is obvious that for some countries with a limited inventory, in particular for countries that are operating only research reactors, the cost of the development of disposal facilities may be very high compared to the benefits from their operation. At present several countries return their research reactor fuel to the country of origin of the fuel if possible, and thus it does not require disposal in the country where it has been used. The Russian Federation and the USA have had agreements to take back such research reactor fuel. There are some countries that have decided to reprocess research reactor fuel in other countries, such as Belgium and Australia, which have made the choice to reprocess in France. The international experience accumulated from research reactor fuel take-back programmes for HEU has been collected and presented in Ref. [31].

7.3. MANAGEMENT OF INTERMEDIATE LEVEL WASTE

Globally the volumes of ILW are small compared to LLW and VLLW, typically around 5% of the total. Many industrial scale methods exist for safe processing, packaging and storage of ILW. Many countries (France, Japan and the UK, for instance) plan to develop an underground disposal facility for ILW co-located with HLW. There are some new combined ILW and LLW disposal facilities in operation: one is under construction (Konrad in Germany) and three are in the regulatory approvals process (Cigéo in France, SFL in Sweden, which is planned for the late 2030s, and OPG's DGR in Canada).

Since the last report, advancements in ILW management include the following:

- In the USA, operations resumed in 2016 at WIPP after incidents in 2014 that caused temporary shutdown [74].
- Also in the USA, the Department of Energy issued an Environmental Impact Statement for the Disposal of Greater-Than-Class C (GTCC) Low-Level Radioactive Waste and GTCC-Like Waste [75]. The preferred alternative is land disposal at generic commercial facilities and/or disposal at the WIPP geological repository. A final decision is pending Congressional action.
- The new date given by the BGE in the draft framework schedule for commissioning of the Konrad mine repository in Germany is the year 2027. Konrad is for all non-heat-generating waste, including long lived ILW, and is under construction.
- Cigéo in France (which will accommodate long lived ILW) is currently in the regulatory approvals stage.
- In the Russian Federation, long lived ILW waste is required to be disposed of in a DGR by law [76].
- The RECUMO project, launched in Belgium, aims to decontaminate current and future highly radioactive residues in cooperation with SCK CEN and the National Institute for Radio Elements.

Special challenges are connected to some categories with relatively larger volumes (e.g. graphite from gas cooled reactors and radium bearing waste from earlier radium production). Although the radioactivity levels in some of the waste can be relatively low, it is composed mainly of long lived radionuclides, such as ^{14}C and ^{226}Ra, together with its decay products and is therefore managed as ILW. The issue is to develop a solution proportionate to the actual hazards of this waste, which is less active than some other ILW waste.

7.4. MANAGEMENT OF LOW LEVEL WASTE

Historically, disposal solutions have been developed first for this category of waste. The classification and safety criteria, in particular for the long term, have been established gradually and have mainly addressed operational waste. There are a number of countries with disposal facilities for LLW; however, due to the planned decommissioning of their nuclear installations, there might be a need for additional capacities in the near future.

Most disposal facilities for LLW are surface or near surface facilities, including relatively shallow depth, underground caverns. The surface or near surface facilities can be found, for example, in Spain, France, and Romania. However, in several countries, such as Sweden, Hungary, Germany and Switzerland, the choice has been made to dispose of this waste in deeper rock caverns.

The development of disposal facilities dedicated to this waste is progressively continuing in the world:

- Several new disposal facilities recently started operation. Examples include facilities in the Russian Federation (Ural region), Hungary (Bátaapáti), the UK (extension of Dounreay near surface disposal facility) and the Republic of Korea (Wolsong).
- Others are under construction or already licensed, for example Konrad in Germany, Stabatiškės in Lithuania and Radiana in Bulgaria.

- Some others are under the regulatory process of approval in Canada (Kincardine DGR and Chalk River Near Surface Disposal Facility), Slovenia (Vbrina-Krško), Romania (Saligny), the Islamic Republic of Iran (Talmesi) and Dessel in Belgium.
- There are also several site selection processes going on, e.g. in Australia, Malaysia and Pakistan.

As existing disposal facilities are considered as valuable assets that are difficult to replace, waste minimization is an important issue in order to expand their lifetime: for the Low Level Waste Repository facility in the UK, the lifetime has been expanded through a successful waste diversion programme (up to 85% of new waste being diverted, resulting in the diversion of some 50 000 m^3 during the period 2008–2016). The operator promoted decontamination and recycling options as well as diversion to VLLW landfills. This is a shared concern by all disposal facility operators. The waste minimization approach is also being applied in other countries (e.g. France and Spain) thanks to the definition of VLLW as a subclass that makes it possible to develop dedicated management and disposal facilities for such waste.

Waste minimization is a growing challenge as decommissioning waste streams are increasing. To face this issue, an expansion of the SFR facility in Sweden to accommodate decommissioning waste is in the licensing stage. In the Republic of Korea, optimization is planned by expanding the existing silo type disposal facility at Wolsong with the addition of a near surface facility for LLW.

The construction of a dedicated disposal facility for LLW in countries with small quantities of waste may be difficult, owing to the relatively high initial fixed cost to site, design, licence and construct a repository. Alternative solutions, such as borehole disposal, may be more practical and cost effective to implement.

7.5. MANAGEMENT OF VERY LOW LEVEL WASTE

According to GSG-1 [3], VLLW is a waste that "does not need a high level of containment and isolation and, therefore, is suitable for disposal in near surface landfill type facilities with limited regulatory control". Some States (e.g. the USA and Spain) consider VLLW as a subclass of LLW and dispose of it in different areas of the same facility. Others have created separate facilities for the two types (e.g. Sweden, France and Lithuania). Others still have licensed appropriately permitted landfill sites to accommodate VLLW (e.g. the UK and the Netherlands). The implementation of disposal facilities dedicated to VLLW is the result of a desire for optimization and to reserve LLW disposal facilities for waste that truly requires the higher level of engineered barriers and isolation provided by them.

The relevance of this category will increase during the dismantling operations of nuclear facilities: large volumes of VLLW are expected in the future as a result of the decommissioning and dismantling programmes of nuclear power plants (see Section 3.2.1). Preparedness to accommodate large quantities of waste for disposal over a fairly short period of time will be important — particularly in countries that foresee an accelerated decommissioning programme for nuclear power plants. In addition, much of the waste produced as a result of dismantling a nuclear facility is different from normal operational waste, as it has a higher proportion of minimally contaminated metals (e.g. equipment) and building rubble. Cost effective disposal solutions that still provide adequate safety for this class of waste have been and are being developed in various countries.

However, despite this relevance, it is not to be forgotten that, as for LLW, disposal facilities for VLLW are rare assets and their availability has to be preserved for as long as possible. There is a risk of early saturation of disposal capacities in some countries, such as France. Therefore, countries are also focusing on alternative solutions, such as the following:

- Waste minimization at the source: characterization may be a significant challenge as the lower the activity, the longer and the more difficult activity assessment is. Reliable industrial assessment is required. Generally, characterization of VLLW is also connected to free release issues, to segregate waste to be managed in this specific radioactive route from exempted waste. These challenges are approached by related R&D, supporting operational experience.

- In several countries, it is possible to partially avoid the production of radioactive waste, by the implementation of recycling of radioactive material. Again, adequate characterization is a major consideration, and there have to be clear regulatory processes. It has been successful in some countries as Spain, the UK or Sweden, illustrating the impact of national policy and regulatory frameworks; however, there are also countries where the recycling of radioactive materials is not permitted.

7.6. MANAGEMENT OF RADIOACTIVE WASTE FROM DECOMMISSIONING

The amount and characteristics of waste and other materials to be generated by a given decommissioning project is related to several factors, including reactor type and design, unit history, decommissioning strategy, safety and environmental regulations and radioactive waste management routes. However, typically decommissioning activities imply the generation of large volumes of materials that need to be properly managed. Only a small fraction of that waste will generally be classified as radioactive waste. The amount to be declared as radioactive waste will vary by country, according to their laws, regulations, practices and available waste management infrastructure. For example, it is estimated that about 104 000 tonnes of materials will be managed throughout the duration of the José Cabrera nuclear power plant dismantling project in Spain. Approximately 4% of this will be classified as radioactive waste.[5]

For those materials from decommissioning that are not considered as radioactive waste, it is common to follow reuse and recycling approaches. The main fraction of residual materials generated during these activities will normally become 'conventional' waste that will be managed through standard industrial waste management routes and outside of nuclear regulatory control. Specific attention is required in the case of non-radioactive waste that poses toxic or chemical hazards (e.g. heavy metals, asbestos).

Radioactive decommissioning waste is mostly similar in terms of radiochemical hazard and risk to the radioactive operational waste. Similar approaches and technology are being used for its treatment and conditioning. However, dedicated waste streams may be generated requiring specific consideration.

Success in decommissioning may be strongly influenced by the suitability of three key elements to complete an integrated approach: a regulatory legal framework, the necessary provisions with regards to the funding and availability of resources, and access to technologies and experience in this field, including the presence of logistical and management solutions for the resulting materials, particularly radioactive waste.

Early planning is needed, prior to the start of actual decommissioning, to enable proper and timely treatment, conditioning and disposal of the radioactive waste from decommissioning. It is important to ensure that 'orphan' waste is not generated during the decommissioning process (i.e. waste with no known safe disposition).

Most operating radioactive waste disposal facilities are for a given class or type of waste, and will accept waste from any source as long as it meets the acceptance criteria for the facility (e.g. it can take waste from either the operation or decommissioning of a nuclear facility). However, there are several cases in which such facilities are licensed for operational waste but not for decommissioning waste (e.g. Finland, Japan, Sweden), and others in which technical modifications, such as licensing of 'larger decommissioning packages', have been needed to enable a more efficient use of the available disposal capacity (e.g. Spain).

Optimization of the available waste management infrastructure and disposal routes have to be considered because of its influence on two main decommissioning schedule drivers: time and cost. Lack of a final waste management route (i.e. disposal) need not be an excuse for postponing decommissioning or preparation for it. However, the implications that may arise from early decommissioning under

---

[5] http://www.enresa.es.

those circumstances need to be carefully considered in order to not jeopardize future necessary actions. Experience from past decommissioning projects shows the importance of optimizing the overall waste management approach, e.g. via a full understanding and use of the waste management hierarchy (see Section 4.8).

While most decommissioning programmes focus on nuclear reactors of various types, all facilities that contain or handle radioactive materials will eventually have to undergo decommissioning and dismantlement. This includes research facilities, waste treatment and conditioning facilities, storage facilities, nuclear application installations, fuel cycle facilities, etc. The scale of these projects is usually smaller than nuclear power plant decommissioning. However, some large and complex facilities, such as spent fuel reprocessing plants, can be equally, if not more so, challenging to undertake. These other types of facilities tend to be much more single event type projects, with little experience to draw from. Consequently, the range of waste resulting from the decommissioning of non-reactor nuclear facilities varies widely both in terms of quantity and radioactivity. Other hazards, such as chemical hazards, also require careful consideration in the planning process of these facilities.

Management of large components constitutes a dedicated point of interest, as this activity is usually particular to decommissioning [77]. Different approaches have been taken to date in response to the peculiarities of individual projects and associated management routes. For example, in some countries, such as the USA, large components can be disposed of in one piece, perhaps after stabilization by filling them with grout. In other countries, such as Germany, the prevailing practice is to cut the large components into smaller segments that will fit inside a 'standard' WP.

7.7. MANAGEMENT OF DISUSED SEALED RADIOACTIVE SOURCES

Most high activity sealed sources are returned to the country where they were manufactured and will thus be included in the waste management system of that country. However, historical waste, orphan sources and sources of lower activity remain a challenge for some States. IAEA Nuclear Energy Series No. NW-T-1.17 [33] provides guidance on methodologies and techniques that could be used to locate, identify and characterize disused, sealed radioactive sources on historical waste sites. The disposal of DSRSs is still a challenge in most countries and new concepts are currently being developed.

One main issue is to create an inventory and a follow-up system in the countries to provide a good knowledge of the sealed sources that are used and who the users are. In addition, a collecting system of disused sealed sources with a (financial) motivation for the user to use this system has to be implemented. There are several international initiatives for securing the safe storage of the disused sealed sources.

Recognizing the need to assist Member States in the safe and effective management of disused sources, the IAEA has focused on the development of a series of publications dealing with the handling, conditioning, storage and disposal of such sources. Guidance on the management of DSRSs, including problems encountered and lessons learned, is included in IAEA Nuclear Energy Series Report NW-T-1.3 [27] to support well informed progress and decisions in managing disused sources. There are several ongoing international initiatives to safely manage DSRSs, and the construction of the first borehole disposal facility for DSRSs is in process.

7.8. SPECIFIC WASTE MANAGEMENT ISSUES

The following subsections discuss a number of specific topical issues of relevance to a range of countries.

7.8.1. Management of waste from nuclear accidents

Two major nuclear accidents in the past decades have resulted in widespread contamination that has required significant remediation efforts — Chornobyl and Fukushima. The accidents and their aftermath are well documented in open literature [78, 79]. Work at both sites is currently ongoing and will likely be continuing for some time to come.

One major achievement on the Chornobyl site was the completion in November 2016 of the largest moveable land based structure ever built (total weight of 36 000 tonnes equipped), a new shelter installed above the original shelter that was emplaced over the reactor after the accident. The next steps now include dismantling unstable structures of the original shelter and then retrieving material contaminated by fuel debris. In addition, there will be some 300 000 m^3 of solid waste to manage. Temporary storage created after the accident in the exclusion zone 30 km around the damaged reactor will also be recovered (1 million m^3). An overall strategy for waste disposal also needs to be implemented. At Chornobyl, good progress has been made in developing permanent solutions within the exclusion zone, but challenges still remain even after several decades.

The Japanese government has formulated a roadmap towards the decommissioning of TEPCO's Fukushima Daiichi Nuclear Power Station and is engaged in process control measures for decommissioning. On the Fukushima site [80], the completion of fuel removal from the spent fuel pool of Unit 4 was achieved in December 2014. Removal of large pieces of rubble on the refuelling floor and spent fuel pool (Unit 3) was completed in 2015. Due to the high dose level, additional measures, including usage of remote control machinery, decontamination and shielding, have to be considered to reduce radiation exposure at units 1 to 3 for the removal of fuel. In 2019 the removal of spent fuel started.

Walls of frozen soil have been installed at Fukushima since March 2016 to prevent infiltration of groundwater in the bottom parts of the reactors. Large amounts of low level tritiated water (more than 900 000 m^3) are stored on-site. The experiences and lessons learned in the cleanup and decommissioning of nuclear facilities in the aftermath of accidents are also presented in IAEA Nuclear Energy Series No. NW-T-2.7 [34].

7.8.2. Management of areas affected by past activities and waste from past activities

In many countries, there is waste from past nuclear activities and from contaminated sites. These include some older facilities, which were developed according to the best practices available at that time, but which may not meet current standards or practices in radioactive waste management and may require remedial actions to ensure ongoing safety. Many of these sites and/or waste are poorly characterized (e.g. radioactivity content and chemical composition) and can exist in substantial quantities (e.g. contaminated soils and liquid waste in tanks) as well as in unconfined areas (e.g. no engineered barriers around them). Many of these sites date back to the very early years of radioactive material processing when there was no full awareness of the hazards. They can also be the result of non-nuclear industries that extract or handle radioactive materials incidental to their primary purpose (e.g. mineral extraction and processing) and that were not subject to nuclear regulatory control. In addition, original records of the waste and its characteristics, how it was managed and/or precisely where it was disposed of may not exist, may have been lost or may otherwise have become unreadable over time.

The location of sites and the assessment of their inventory and potential hazards may therefore be difficult. The management and disposal of this waste is challenging, both technically and financially, due to the poor characterization, the often widespread geographical area affected and the lack of a definitively 'responsible party' still in existence (e.g. the previous owner or operator of the site has gone bankrupt or otherwise ceased to exist) [33]. In these cases, the responsibility for cleanup generally falls to the State by default.

A variety of techniques, ranging from complete recovery and remediation to in situ disposal, have been implemented. Each case is unique and requires a thorough understanding and evaluation of the consequences of the various options (e.g. removal versus leaving it in place). Remediation programmes are ongoing in a number of countries. For example, in the Russian Federation, a total of 200 ha of contaminated territories were remediated in the period from 2014 to 2016, including mining and milling production sites (Stavropol territory, Kurgan region, Zabaikalsk territory); sites of production enterprises and R&D sites (Moscow region, Vladimir region, Chelyabinsk region); and sites used for peaceful nuclear explosions (Ivanovo region, Tyumen region, Perm territory). Several remediation projects have also been completed in Canada, Germany, the UK and the USA in recent years. Efforts are still under way in these countries, as well as several others, to clean up past legacies.

Often, due to financial and other constraints, the management options are based on finding a safe and optimized solution with respect to radiation protection and worker and public safety [81], possibly with ongoing monitoring, restrictions of use or other institutional controls, but without pretending to necessarily achieve long term safety.

A particular challenge is the treatment of the huge amounts of stored liquid radioactive legacy waste in the Russian Federation and the USA (see Section 6.4.2), mainly ILW and HLW resulting from past military activities. In the USA, multiple facilities have been constructed and operated to process liquids and convert them to a stable solid form. These include facilities at the West Valley Demonstration Project (operated from 1996 to 2002; now demolished); Savannah River site (currently in operation); and Hanford (in construction). Before the treatment is achieved the liquids will have to be kept in safe storage conditions. In the Russian Federation, projects have been recently completed to improve the storage reservoirs at several locations, including the Siberian Chemical Combine in the Tomsk region, Mining and Chemical Combine at Zheleznogorsk and Mayak Production Association.

7.8.3. Research and development in spent fuel and radioactive waste management

Ongoing research and development (R&D) is an important and integral part of most radioactive waste and spent fuel management programmes. Nuclear power countries have extensive programmes for R&D in spent fuel and radioactive waste management. R&D is carried out by a variety of entities, including facility operators, regulators and technical support organizations, as well as by independent organizations, such as universities and research institutes. The specific goals of the R&D can vary from basic science fundamentals to applied research for developing specific technical solutions. Much of the current and proposed research is centred on the development and demonstration of technical equipment required for repository construction and operation (e.g. construction methods, waste handling/emplacement equipment, tunnel sealing, etc.). Collaborative work is also ongoing in areas such as ageing management, predisposal management, high burnup issues, partitioning and transmutation, which could have an impact on the development of advanced fuel cycles, as well as the types and quantities of waste to be disposed of. Development and potential deployment of new reactor types, such as small modular reactors, will also affect the need for and type of fuel cycle and waste management facilities in countries wishing to deploy such technology.

Most countries also have R&D programmes in place to predict future performance and take any corrective action required. In extreme cases, such as the post-accident recovery at Fukushima, specific technology is under development to recover fuel debris from the damaged cores.

There is ongoing R&D to reduce the volume and potential hazard of HLW, for example separation of actinides would help to reduce the hazards related to management of spent fuel, especially long term storage or disposal. Similar effects could also be achieved though transmutation. For example, an advanced fuel cycle that includes partitioning and transmutation in addition to the reuse of uranium/plutonium, e.g. in Generation IV fast reactors or accelerator-driven systems, could provide benefits to geological disposal by reducing the radiotoxicity and thermal output of the final waste inventory, thereby positively impacting the required footprint of such a repository. However, it is necessary to note that, since such a fuel cycle is still in the R&D phase, the actual benefits for the fuel cycle in general and geological disposal in particular are difficult to estimate. There is R&D related to different aspects of small modular reactors and Generation IV reactors. For example, Belgium has approved the Multipurpose Hybrid Research Reactor for High-tech Applications (Myrrha) project, for which construction is expected to begin in 2026.

As stated previously, the length of the storage period could be many decades, but there are regulatory provisions and technical measures in place to ensure continued safe storage over the longer term. A future challenge is related to the handling of the spent fuel and its integrity after several decades of storage, especially in dry storage conditions where it is not possible to directly observe the fuel since it is sealed/bolted into casks or canisters.

Sharing knowledge and experiences needs to be encouraged among the countries under the auspices of international organizations such as the IAEA, OECD/NEA, EC and WNA. Some examples are provided below.

The IAEA has facilitated three Coordinated Research Projects (CRPs) on spent fuel storage issues:

- CRP T13014 on Demonstrating Performance of Spent Fuel and Related Storage Systems during Very Long Term Storage (DEMO) ran from 2012 to June 2016. The project gathered 16 partners from 11 Member States. It aimed to improve the nuclear power community's technical basis for LWR spent fuel management licences as dry storage durations extended. The work conducted within this CRP provided data to partially close many of the data gaps identified, and enabled a dynamic network of experts working on demonstrating the long term performance of spent fuel to be established to share technical information [82].
- CRP T13016 on Spent Fuel Performance Assessment and Research, phase IV (SPAR-IV), ran from February 2016 until 2020 and gathered 11 partners from ten Member States. It gathered operational experience and research results on the behaviour of spent fuel (from all types of power reactors) in wet and dry storage. Within this scope, the IAEA has since 1981 conducted three CRPs on the Behaviour of Spent Fuel Assemblies in Storage (BEFAST I, II and III) and three previous phases of SPAR CRPs, gathering 50-plus years of experience in spent fuel wet storage and 30-plus years of experience in spent fuel dry storage operation [72, 83, 84].
- CRP T21028 on Ageing Management Programmes for Dry Storage Systems (AMP), ran from October 2016and has so far gathered six partners from four Member States. It aimed at collecting and sharing up-to-date R&D on structures, systems and components, monitoring, inspection and surveillance programmes in support of spent fuel dry storage. It also looked at how this information is used in licence or safety justification renewal and collated experiences in developing ageing management programmes for spent dry storage systems.

Additionally, there is a current CRP to develop a standardized framework for the borehole disposal of DSRSs and small amounts of LLW and ILW. The outcome of that project will provide Member States with a package of essential materials for the development of the borehole disposal and make this disposal solution more readily implementable.

The OECD/NEA has performed the following studies:

- An evaluation of the economic aspects of the back end of the fuel cycle. It mainly concludes that, looking at the overall fuel cycle costs, the cost of the open cycle is comparable to that of the closed cycle [42].
- An EGIRM was established, mainly to create a methodology that would help provide a better understanding of the global picture of spent fuel and radioactive waste management [70].

In 2014, the US Department of Energy published the results of the final tests on spent fuel behaviour in dry storage with a burnup above 45 GWd/tU. These tests were performed by EPRI [85].

The WNA has created a dedicated Working Group on the sustainable development of used fuel management, with the main objective of gathering the views of the nuclear industry and stakeholders (including newcomers) on the back end of the fuel cycle. The working group considers how the industry can best respond to these needs, as well as explaining how effective spent fuel management contributes to the sustainability of nuclear energy and supports its development and implementation.

Because of the very long time frames covered by the safety cases for radioactive waste management and spent fuel management, the safety cases are generally based on mathematical modelling and simulation and/or comparison to suitable natural analogues, rather than by direct observation of repository performance. In order to develop, test and calibrate the models, R&D is generally required to establish model parameters and boundary conditions (e.g. diffusion rates, radionuclide transport processes and corrosion mechanisms under repository conditions, design parameters such as rock strength and other properties, etc.). In the shorter term, R&D may also be required to demonstrate the various technologies required to construct and operate a disposal facility, create optimized waste forms, develop technical specifications and waste acceptance criteria, etc. The exact requirements for R&D will vary by country and by waste type and their planned management routes. Often, a certain amount of initial R&D is required in order to be able to decide the appropriate management route.

Some examples of current R&D in the field of long term radioactive waste management include:

- Country-specific R&D: very active R&D programmes in several countries (e.g. Sweden, France, the UK, Canada, Switzerland, the USA).
  — These normally focus on demonstrations of equipment and technology, operation of URLs, improving the science and understanding of issues related to long term safety (e.g. corrosion of containers, behaviour of engineered barrier materials under repository conditions, geosphere characterization, etc.)
  — Some are individual country work, while some are collaborative between several countries. There is a move towards collaboration, since costs to develop, operate and maintain facilities are high. Most of the work is common to a number of programmes anyway, so there is a cost benefit to collaborating and leveraging funding.
- The framework programme in the management and disposal of radioactive waste under the Euratom Horizon 2020 programme.
- The OECD/NEA NI2050 programme on waste and decommissioning — waste related issues are recognized as important for sustaining the nuclear power industry into the future.
- IAEA coordinated projects — looking at specific waste types considered to be problematic for disposal, e.g. irradiated graphite (the GRAPA (Graphite Processing Approaches) project) and waste from next generation advanced reactor systems (the WIRAF (Wastes from Innovative Reactors and Fuel cycles) project).

On the EU side, following implementation of the Euratom Waste Directive, most Member States provided their national programmes in 2015 and report according to the Directive. A Joint Programme of common research on radioactive waste management, including deep geological disposal (EURAD) started in 2019. In addition, there are multinational consortia working on large scale projects. The projects are examining large, difficult to answer questions in a comprehensive fashion. EU based projects such as THERAMIN (Thermal treatment for radioactive waste minimization and hazard reduction) and CHANCE (Characterization of conditioned nuclear waste for its safe disposal in Europe) are still active, while others such as DOPAS (Full scale Demonstration of Plugs and Seals), JOPRAD (Joint Programme on Radioactive Waste Disposal) and FORGE (Fate of Repository Gases) finished in recent years. All concern various aspects of radioactive waste management encompassing many different European countries and organizations, including State regulators, universities, private corporations and State research agencies, as well as some non-European countries and institutions from Asia and the Americas.

In addition to the formal research programmes, the IAEA, OECD/NEA and WNA all host and support peer-to-peer networks and expert working groups on a range of topics related to radioactive waste management, spent fuel, fuel cycles and decommissioning. These networks and working groups facilitate the exchange of information and experience among their members.

There are different projects and networks to share knowledge as well as infrastructure. For example, among others, the IAEA's International Centres based on Research Reactors (ICERRs) scheme is intended to help IAEA Member States gain timely access to relevant nuclear infrastructure based on research reactors to achieve their capacity building and R&D objectives.

8. TRENDS

This is the second Status and Trends publication of this series, the main purpose of which is to highlight both status and noted trends in the management of spent fuel and radioactive waste. This section of the report is divided into two subsections. First, general trends are presented, and second, a review is given of the challenges that were identified in the first report [1]. These are summarized, and updates on their current progress and status are provided.

8.1. GENERAL TRENDS

In this section general trends are presented based on information provided in the National Profiles, in discussions during the Fifth (2015) [86] and Sixth Review Meeting (2018) [7] of the Contracting Parties to the Joint Convention, and in the first report of the Euratom Waste Directive [4]. Input from the meetings of IAEA technical working groups (the Radioactive Waste Technical Committee and the Technical Working Group on Nuclear Fuel Cycle Options and Spent Fuel Management) was used as well. This section outlines some achievements, emerging issues and challenges that have been identified since the publication of the first Status and Trends report. For convenience, they are grouped into a number of functional areas or themes.

8.1.1. Policy and strategy

Radioactive waste, spent fuel management, and decommissioning policies and strategies are generally developed and set by government ministries or agencies. Defining long term aims in the policies and strategies usually helps to ensure successful and optimized implementation. Many countries have successfully established policies and strategies that have been stable over several decades and good progress has been made towards their implementation. An overview can be found in Annex 2 and the National Profiles. Maintaining a stable policy and strategy over time can be a challenge for various reasons. For example, the decision made in Germany to phase out nuclear energy will also affect its spent fuel and radioactive waste management. In other cases, a national policy may have evolved around practices that have been established for decades, but these approaches or strategies might no longer be preferred given the circumstances of today. This means that revisiting the decisions made in the past is also important (e.g. in the case of Dounreay near surface disposal facility in the UK). Spent fuel management is a long term commitment and the strategies adopted for managing the spent fuel produced by power reactors need to keep some flexibility to enable potential changes in policy decisions.

More and more countries are developing, or at least considering, integrated and holistic approaches to spent fuel and radioactive waste management. This has benefits including more effective use of resources, such as disposal space, as well as reducing overall costs. For nuclear newcomer countries, this is a very important consideration to ensure that spent fuel management and radioactive waste management systems are established in an optimal way right from the beginning. More attention and corresponding effort are put into ensuring consistency and safety in predisposal and disposal activities in radioactive waste management. Similar efforts can also be seen related to spent fuel storage and disposal activities.

Some overall tendencies can be seen in national decisions in spent fuel and radioactive waste management. Often countries opt for a single solution for management of the same type of waste or for all waste producers in a country. The siting process for disposal sites is usually undertaken by national waste management organizations, which are usually independent of the waste producers/owners. The siting process is often a broad community consent based engagement process, in which a lot of factors are taken into account. There are safety related questions, but additionally the selection process includes estimation of the environmental impact, transportation issues, cost, etc.

8.1.2. Facilities

LLW and VLLW management and disposal has been safely implemented for many years in many countries and there are a number of the disposal facilities available. There is, however, a need in several Member States for expanding disposal facilities in the very near future. A growing number of programmes are developing measures to increase available storage and disposal capacities. Several countries are currently constructing new disposal facilities for LLW, such as China, Germany and the Russian Federation, while others are currently engaged in the regulatory approvals process for their first repositories, e.g. Belgium and Canada. Some countries, such as the UK, have been successful in diverting waste from their national LLW repository through a combination of recycling of material and diverting very low activity waste to other facilities, such as designated industrial landfills. Other countries, such as France and Spain, have built specially designed facilities for VLLW and in doing so have lowered the amounts of waste going to their national LLW repositories.

The prolonged storage periods for spent fuel caused by lack of disposal and/or lack of sufficient reprocessing capacity have resulted in the need for additional storage capacity. For example, where there is no additional storage capacity available on NPP sites, new storage facilities are planned (e.g. in the Republic of Korea). Usually new storage facilities comprise dry storage outside of the reactor buildings, but there are also examples of additional wet storage being built.

A particular issue related to storage and disposal space for radioactive waste is for post-accident situations. Both Ukraine (Chornobyl) and Japan (Fukushima Daiichi) are facing challenges developing the facilities required to safely manage the large volumes of accident related waste.

There is a clear trend that available radioactive waste predisposal management facilities are used by several countries, and on an international scale. The waste from other countries is sent for processing and will be sent back to the original country after treatment. There are existing facilities that have been providing waste treatment services to domestic and international customers for 40 years.

Another challenge is related to the availability of disposal, especially for DSRSs, in countries with small programmes and/or little existing infrastructure. Borehole technology has been considered a suitable disposal option [87], especially for countries with small inventories. The implementation of borehole disposal for DSRSs is planned in Malaysia.

Taking into account the human and financial resources needed for the implementation of disposal facilities, some countries would consider the possibility of sharing efforts towards the implementation of a DGR. This raises the challenging question of political and public acceptance issues across multiple countries [28]. One of the main challenges is to find an informed and willing host community with suitable geology that will accept the disposal of these materials within its jurisdiction from one or more countries.

The issue facing several countries is the stress that has been put on spent fuel and waste management infrastructure by decisions to phase out nuclear power and/or early shutdowns of reactors. These phase out/early shutdown decisions can be political (e.g. Germany) or economic (e.g. USA). In either case, the sudden early influx of decommissioning waste may overwhelm the existing capabilities for managing it. In addition, early shutdown often results in a shortfall in decommissioning funds, which are normally collected over the full operational life of the reactor.

8.1.3. Technology/innovation

Continuous efforts are made to develop new tools and technology in response to identified waste management issues. For example, with the growing interest in segregating VLLW and, in some countries, clearing material from the radioactive waste stream, there is a need for faster and more accurate characterization equipment and techniques. This is especially true for the non-gamma-emitting 'difficult to measure' radionuclides that are often the limiting factors in determining the suitability of waste to be put into a lower classification. The characterization is needed both for the initial classification of the waste into the appropriate stream and for compliance monitoring purposes.

Several 'problematic' waste streams require the development and testing of new treatment and conditioning technologies to ensure their safe disposal. For example, graphite waste that originates from dismantling the core of some reactor types requires treatment and conditioning prior to disposal. This type of waste is becoming more prominent as graphite containing reactors are being decommissioned. International collaborative projects, such as the IAEA's GRAPA (Irradiated Graphite Processing Approaches), are currently addressing this issue. Other problematic waste requiring specialized solutions includes sodium and sodium wetted materials from sodium cooled reactors, as well as lead and lead-wetted materials from lead cooled reactors. Again, these issues are becoming more prominent as several of these types of reactors are being decommissioned.

An important aspect of innovation and the development of solutions for problematic and possible future (from reactors under the development, including small modular reactors) waste streams is international collaboration and sharing of information. This is facilitated by the various international organizations (e.g. the IAEA, OECD/NEA and WNA) through their sponsorship of working groups, technical meetings, CRPs and on-line discussion forums.

8.1.4. Governance

A key component of radioactive waste and spent fuel management programmes is 'continuous improvement'. There are several review services available for countries and these services are used more often than in the past, as there is increased recognition of the expert peer reviews organized by international organizations such as the IAEA and the OECD/NEA. In fact, such peer reviews are required for EU Member States under the Euratom Waste Directive [4].

The IAEA's Integrated Review Service for Radioactive Waste and Spent Fuel Management, Decommissioning and Remediation (ARTEMIS) is an integrated expert peer review service for radioactive waste and spent fuel management, decommissioning and remediation programmes. This service is intended for facility operators and organizations responsible for radioactive waste management, as well as for regulators, national policy makers and other decision makers. Between launching the service in 2016 and the end of 2019, 12 review missions were conducted.

For newcomers in nuclear energy there is, for example, the IAEA Integrated Nuclear Infrastructure Review (INIR), which is a holistic peer review to assist countries in assessing the status of their national infrastructure for the introduction of nuclear power. The review covers the comprehensive infrastructure required for developing a safe, secure and sustainable nuclear power programme. The topics covered also include nuclear fuel cycle and radioactive waste management. Since the start of the service in 2009, 26 such review missions for the different phases of developing a nuclear power programme have been conducted.

Another key aspect is the growing interest of the public and other stakeholders in nuclear activities in general and radioactive waste and spent fuel management in particular, leading to an increased awareness of the role played by society in effective implementation of radioactive waste and spent fuel management programmes and projects, in particular related to disposal. This has led to many countries adopting a more open and transparent system for communicating with the public and other stakeholders, as well as to the development of more participatory regulatory review processes. The challenge is for the decision makers to balance technical considerations with sociopolitical and economic considerations.

8.1.5. Funding and financing aspects

The successful implementation of a radioactive waste and spent fuel management programme (and/or nuclear decommissioning programme) requires adequate and stable funding, often over many decades or longer. While all countries with nuclear power programmes have established specialized funds to support these programmes and ensure that undue burden is not passed on to future generations, the challenge is to ensure that the funds remain viable, secure and available over the long time periods required. The situation in each country will be different, and various mechanisms have been established in different countries to ensure the future adequacy of the fund. One important aspect common to many countries is the requirement for periodic review and assessment of the fund and its management. This includes a review of both the cost estimates and the investment strategy for the fund, as well as any economic parameters used in calculating the adequacy of the fund (e.g. inflation rates, interest rates, financial discount factors, taxation implications, etc.). It is clear that increased social and political interest will help to guarantee the adequacy and sufficiency of the funds available for the safe management of spent fuel and radioactive waste.

8.1.6. Regulatory framework

There are internationally agreed standards for the safe management of spent fuel and radioactive waste, such as the Joint Convention [2]. In order to secure the safety of the public and workers, a regulatory framework has been put in place in the majority of countries. Examples can be found in the National Profiles. There has to be collaborative effort between the implementer and the authorities for the licensing of spent fuel or radioactive waste management activities and facilities, although the regulator has to remain independent from the implementer. Experience has shown that licensing processes for radioactive management facilities, and in particular for geological disposal facilities, are complex and often carried out over very long timeframes. They are also generally limited in number in any one country (i.e. it is unlikely that any single country will have more than a few repositories for radioactive waste or spent fuel). As such, a major challenge is for both the regulator and the implementer to build up and maintain the expertise and competencies required for licensing, constructing and operating a disposal facility. Because of this 'one-off' nature, early interaction between the implementer, the nuclear regulator and perhaps other regulatory bodies can have benefits in clarifying the licensing process. An integrated approach to the regulatory process (e.g. considering all aspects at once: nuclear safety, environmental issues, health and safety, mining law, etc.) maintains clarity. In practice, regulatory interaction could be achieved by a multistep licensing process. The Contracting Parties at the Sixth Review Meeting of the Joint Convention [7] indicated that the feedback and information from countries with experience with such licensing processes would be of considerable benefit for all countries currently engaged in or contemplating licensing disposal facilities.

8.1.7. Knowledge system sustainability

It has long been recognized that the workforce in the nuclear industry is ageing, and a considerable fraction of workers are at or near retirement age. A challenge facing the industry is how to attract enough trained and skilled staff in all disciplines, and transfer knowledge from the current generation to the new generation of workers. This issue is faced by all organizations involved, including implementers, regulators, technical support organizations, researchers, support industries, etc.

While the responsibility for the safety of radioactive waste management is primarily that of the waste generator, national radioactive waste management programmes require a degree of national capability to be in place. The availability of the scientific, engineering and legal skills necessary to implement and regulate national programmes requires educational and training provisions to be in place. Expertise in specialist scientific disciplines needs to be available and research capabilities are required. Bearing in mind the timeframes associated with the development, operation and closure of radioactive waste management facilities, in particular for storage and disposal facilities, this matter of human resources is of fundamental concern to all countries, as is an understanding of the necessary skills base and experience in maintaining such skills. The availability of sufficient financial resources remains a challenge for many spent fuel and radioactive waste management programmes, and is particularly important for the back end activities of decommissioning and disposal. The availability and feedback of knowledge and experience on costing and financial provision for both back end activities and legacy situations is of considerable value. International cooperation, either between different organizations or under the umbrella of international organizations, has an enhanced role in sharing knowledge and in securing the safe management of spent fuel and radioactive waste.

8.2. REVIEW OF PREVIOUSLY IDENTIFIED CHALLENGES

The first edition of this publication [1] detailed a number of challenges related to the management of spent fuel and/or radioactive waste. These are summarized below, along with updates on their current progress and status.

8.2.1. Public acceptance of spent fuel and radioactive waste management remains a challenge in most countries

Involving interested parties in every stage of the life cycle of nuclear facilities is essential to enhance mutual trust on issues related to nuclear energy production. The process of developing and building DGRs is not purely scientific or technical, but also requires a phased consultation process engaging a range of stakeholders [48]. The same principles are to be used during environmental remediation programmes [35] or decommissioning [32]. It is important to recognize that beyond the groups traditionally involved in the decision making process, such as the nuclear industry, scientific bodies and relevant national and local governmental institutions, the concept of stakeholders also includes the media, the public, local communities and non-governmental organizations [30].

Informed participation of stakeholders in the development of a nuclear facility (including reprocessing, storage and disposal facilities) is crucial. Consent is not meaningful unless it is informed. This means that potential host communities need to receive adequate information, financial and technical resources to enable effective participation and provide for informed decision making. Meaningful public participation in an environmental impact assessment ensures an open, balanced process and strengthens the quality and credibility of a project's review.

An effective strategy of local stakeholder involvement encompasses more than applying rules and guidelines. Each group of local stakeholders has its own interests and influence, and there is a diversity of national circumstances across countries and even across local communities. This means that a tailor-made approach may well be justified and required. However, there are always some general lessons learned that can be useful for the design and implementation of spent fuel and radioactive waste management programmes. The experience of many countries has shown that building and maintaining public trust is a long process, and it is vital to keep local people involved during a facility's operation. This means active participation by members of civil society in the technical monitoring of activities. In addition, hosting public tours of the facilities, open houses and other events can help to engage the public.

A variety of communication methods are actively being used by various organizations involved in spent fuel and radioactive waste management. Development of a comprehensive safety case in easily understandable terms is an important mechanism for communicating the results of the safety assessment and the overall safety argument for the facility to the regulatory authorities. The type of strategy and methods to be used for communication and stakeholder involvement need to be established in an appropriate manner that reflects the specific national societal, political and institutional situation, as well as the specific needs and concerns of the stakeholder group.

There are several international initiatives for sharing experiences, and both the IAEA and OECD/NEA have ongoing work programmes related to building and maintaining stakeholder confidence, e.g. the IAEA's Nuclear Communicator's Toolbox[6].

---

[6] https://www.iaea.org/resources/nuclear-communicators-toolbox.

8.2.2. Funding for waste management and decommissioning activities still remains a challenge in a number of countries

There is global understanding that successful implementation of a radioactive waste management, spent fuel management and/or nuclear decommissioning programme requires adequate and stable long term funding. The situation is improving in many countries, and the requirements of the Joint Convention and the Euratom Waste Directive have had positive effects. Countries with nuclear power programmes have established specialized funds or mechanisms to finance these programmes and ensure that undue burden is not passed on to future generations. There are, however, still some challenges remaining, such as how to ensure that the funds remain sufficient, viable, secure and available over the long time periods required. Different options have been implemented by different countries, but one important aspect common to many countries is the requirement for periodic review and assessment of the fund and its management. The estimation and the review of the cost estimates might be especially challenging for the countries, which are planning/building their first NPP, as the implementation and real costs will accrue in the decades from now. The main remaining challenges are related to large historical or legacy liability issues, which are very costly to resolve. This requires considerable and long term funding, which can be difficult to secure taking into account that State budgets are approved annually and there are always competing priorities. As can be seen from Annex 4, most countries have financing schemes and funding mechanisms in place, but there are still a number of countries that struggle with the legacy liability.

8.2.3. Safe retrieval and handling of spent fuel after a long period of storage

Ageing management of storage facilities and containers of spent fuel is a growing challenge. As the implementation of the disposal options for spent fuel and HLW is postponed in many countries, the storage periods lengthen, often by decades. A number of recent and ongoing collaborative research programmes have addressed ageing management issues, especially related to spent fuel storage. Examples are listed in Section 7.8.3.

A related issue is the challenge associated with the transportation of higher-burnup spent fuel and damaged fuel, as fuel and cladding integrity is a matter of concern in higher-burnup fuel. However, there are also a number of initiatives being undertaken to enhance the safety of spent fuel management, taking into account the economic aspects. Examples include progress in higher-burnup reprocessing opportunities, proliferation resistant reprocessing processes and the potential reduction of HLW radiotoxicity and volumes [88].

8.2.4. Decommissioning planning

More than half of the power reactors currently operating in the world are more than 30 years old (Fig. 4). Therefore, the decommissioning of nuclear facilities is gaining in importance as an increasing number of reactors and related facilities are being permanently shut down, or will be in the near future. This means that more and more decommissioning activities will be ongoing in different countries. An increase in decommissioning activities results in greater challenges for planning, dismantling, funding and, in particular, waste disposal. In some cases, a lack of disposal sites blocks or delays the start of dismantling. A streamlined waste management approach and disposal infrastructure are becoming increasingly important. A major decommissioning issue concerning spent fuel is related to the need for additional AFR storage capacity in order to be able to defuel the core and the at-reactor pool.

The decommissioning and dismantling of reactors damaged in accidents represents a particular challenge. Japan has established the International Research Institute for Nuclear Decommissioning to help support decommissioning of the accident-damaged Fukushima reactors.

In recognition of the increasing importance of decommissioning, both the IAEA and OECD/NEA have recently established dedicated technical teams to support decommissioning and environmental restoration programmes for their Member States. There are also different platforms and e-tools for sharing the experience available, e.g. the International Decommissioning Wiki. There are international projects covering different decommissioning-related aspects, such as management of risks during decommissioning, costing of decommissioning [89, 90], defining the end-states of decommissioning, etc.

8.2.5. Management of disused sealed radioactive sources

Sealed sources are used in practically every country for various industrial and medical practices. For many countries, these are the only radioactive materials to be handled, and they require storage and eventually disposal. Many countries have made progress with respect to the management of DSRSs. There are a number of countries that have defined in their legislation the requirements to return the used sources to the producer/supplier, establish tracking systems, etc. The disposal of DSRSs is still a challenge in many countries, but several international initiatives exist to address this. The IAEA BOSS (borehole disposal of disused sealed sources) programme [87] is focusing on supporting Malaysia and Ghana. Currently Malaysia is in the process of implementing the first borehole disposal for DSRSs. There is also an ongoing IAEA CRP to develop a standardized framework for the borehole disposal of DSRSs and small amounts of LLW and ILW.

In an effort to scale up the safe and secure management of DSRSs, the IAEA has introduced the new concept of Qualified Technical Centres. The idea behind this initiative is to increase worldwide capability to manage DSRs by encouraging countries with well-equipped centres and trained personnel to provide technical services for the management of DSRs, within their countries and regionally. A Mobile Tool Kit, which is a mobile set of equipment needed for the conditioning of DSRs, has been prepared and will be available for countries that require it.

8.2.6. Management of legacy sites and waste

Legacy sites can be found in many countries. The main challenges with legacy sites, as well as contaminated site waste, are related to incomplete information and typically large quantities of waste. A lot of legacy sites date back several decades, and original records of the waste and how it was managed might be incomplete or lost. However, a great deal of progress has been made, and many old sites have been remediated and the conditioned waste stored or disposed of safely.

Some countries with large legacy issues are making progress in dealing with them through adequately funded programmes, definitive schedules and investments in the necessary infrastructure, such as France, the Russian Federation, the UK and the USA. However, the human, financial and technical resources required are significant and pose a severe challenge in less economically developed countries.

The trend in many countries is towards 'optimized' remediation, taking into account the actual risks and appropriate radiation protection needs along with local regulatory requirements and stakeholder expectations. This implies that complete removal of contaminated material may not be required, or even justified, in some cases. Further guidance can be found in e.g. Ref. [81].

9. CONCLUSIONS

This publication provides an overview of the status of spent fuel and radioactive waste management globally, and presents global estimates of the amounts of residual radioactive material accumulated by nuclear activities. Significant progress has been achieved, particularly during the last 10–20 years, in the treatment, conditioning and storage of spent fuel and radioactive waste and in developing national inventories. Radioactive waste and spent fuel are safely managed all over the world, and great progress has been made on a global scale in providing more transparent and credible information.

There has also been progress in emplacing certain types of radioactive waste, as a majority of VLL and LL waste has been disposed of using well known solutions. However, spent fuel, HLW and the major part of ILW remain in safe storage as disposal solutions are delayed. Research undertaken over several decades has progressed to the point that at least three DGRs are to start operation in the next ten years. All three are located in Europe. For countries with small inventories, the development of disposal facilities can be a challenge, not only for spent fuel or HLW, and for this reason there has been more discussion about the possible sharing of efforts.

The data presented in Section 6 provide a comprehensive overview and a best available estimate of the amounts of spent fuel and radioactive waste that currently exist in the world. The main source of information used for inventories and forecasts is the National Profiles on the web site accompanying this publication. Information about inventories in States that did not submit a report is taken from the reports to the Joint Convention and other publicly available sources.

Worldwide, there is an estimated 263 000 t HM of spent fuel in storage, and 127 000 t HM of spent fuel has been reprocessed. Currently, about 7000 t HM are discharged every year. There is a decline of fuel management by reprocessing; as the UK is in the process of stopping reprocessing activities. This trend may change in the future with the development of nuclear energy and associated fuel cycle facilities in countries such as China. Therefore, special attention needs to be paid to setting up storage capacities and keeping them safe over time, the storage period generally being several decades. This also applies for HLW packages from reprocessing. Some countries have decided to use centralized storage, while others store spent fuel on-site. About 70% of spent fuel is currently stored in pools, but a new development is that most of the additional spent fuel storage facilities are dry storage.

The overall worldwide generated volume of solid radioactive waste at the end of 2016 was about 38 million m^3, with a mean increase of about 1 million m^3 per year. Most of this waste (93%) is VLLW or LLW, and 30 million m^3 of solid waste have already been disposed of (since 2013, about 650 000 m^3 have been disposed of per year), while a further 7.2 million m^3 (19%) are in storage awaiting final disposal.

Available data are sufficient to provide a clear representation of the global situation in terms of the overall challenge represented by the radioactive waste that currently exists, and to provide an indication of the challenges that will arise in the future as facilities still in operation, or planned, come to the end of their useful lives. There remain some uncertainties about total global amounts of radioactive waste, as information about the spent fuel and radioactive waste inventories of all countries is not available. Additional uncertainty in the aggregated data for individual waste classes results from the need to convert data presented according to national classification systems into a common system based on the waste classification scheme of GSG-1 [3]. Finally, uncertainty also arises from the need to present waste quantities according to the anticipated 'as disposed' volumes.

Due to the age of many nuclear power plants, decommissioning will become a more and more important activity. The first generation of nuclear power plants are reaching the end of their design lives. This will be reinforced by changes in nuclear policy in some countries that require the early shutdown of reactors. In addition, many countries are now making concerted efforts to clean up past nuclear legacy sites. Therefore, the availability of disposal routes, in particular for VLLW, will become more and more important. In parallel, as disposal of any waste is not the most preferred option and as disposal facilities for radioactive waste and/or spent fuel are rare assets, there is the challenge of promoting and implementing waste minimization techniques or recycling after decontamination, in an overall optimization approach.

The availability of disposal routes is also highly important for the management of waste generated by nuclear accidents. In Ukraine and in Japan, significant progress has been made towards the remediation of the damaged nuclear facilities (e.g. implementation of shelters, retrieval of spent fuel, etc.) and the surrounding areas (decontamination). These works have generated large amounts of liquid and solid waste that have to be managed with a long term perspective.

DSRSs do not present a challenge with respect to their volumes. However, they present a safety issue due to their quantity (several million) and because they are widely distributed around the world for industrial uses. Some countries already have a system of follow-up of sealed sources in order to enable and motivate their recovery after use. In 2018 the IAEA approved the Guidance on the Management of Disused

Radioactive Sources [10]. International cooperation is needed to develop local management solutions in countries (including, for instance, borehole disposal) or recovery of sources by the countries where they were manufactured.

REFERENCES

[1] INTERNATIONAL ATOMIC ENERGY AGENCY, Status and Trends in Spent Fuel and Radioactive Waste Management, Nuclear Energy Series No. NW-T-1.14, IAEA, Vienna (2018).

[2] Joint Convention on the Safety of Spent Fuel Management and on the Safety of Radioactive Waste Management, INFCIRC/546, IAEA, Vienna (1997).

[3] INTERNATIONAL ATOMIC ENERGY AGENCY, Classification of Radioactive Waste, IAEA Safety Standards Series No. GSG-1, IAEA, Vienna (2009).

[4] EUROPEAN UNION, Council Directive 2011/70/Euratom of July 2011, establishing a Community framework for the responsible and safe management of spent fuel and radioactive waste, OJEU **L199** (2011).

[5] EUROPEAN COMMISSION, Report from the Commission to the Council and the European Parliament on Progress of Implementation of Council Directive 2011/70/Euratom and an Inventory of Radioactive Waste and Spent Fuel Present in the Community's Territory and the Future Prospects, European Commission, Brussels (2019).

[6] OECD NUCLEAR ENERGY AGENCY, Nuclear Energy Data 2016, NEA No. 7300, OECD, Paris (2016).

[7] INTERNATIONAL ATOMIC ENERGY AGENCY, Sixth Review Meeting of the Contracting Parties to the Joint Convention on the Safety of Spent Fuel Management and on the Safety of Radioactive Waste Management, Final Summary Report, JC/RM6/04/Rev. 2, IAEA, Vienna (2018).

[8] INTERNATIONAL ATOMIC ENERGY AGENCY, Code of Conduct on the Safety and Security of Radioactive Sources, IAEA, Vienna (2004).

[9] INTERNATIONAL ATOMIC ENERGY AGENCY, Guidance on the Import and Export of Radioactive Sources, IAEA, Vienna (2012).

[10] INTERNATIONAL ATOMIC ENERGY AGENCY, Guidance on the Management of Disused Radioactive Sources, IAEA, Vienna (2018).

[11] INTERNATIONAL ATOMIC ENERGY AGENCY, Radiation Protection and Safety of Radiation Sources: International Basic Safety Standards, IAEA Safety Standards Series No. GSR Part 3, IAEA, Vienna (2014).

[12] INTERNATIONAL ATOMIC ENERGY AGENCY, Predisposal Management of Radioactive Waste, IAEA Safety Standards Series No. GSR Part 5, IAEA, Vienna (2009).

[13] INTERNATIONAL ATOMIC ENERGY AGENCY, Decommissioning of Facilities, IAEA Safety Standards Series No. GSR Part 6, IAEA, Vienna (2014).

[14] INTERNATIONAL ATOMIC ENERGY AGENCY, Safety of Nuclear Fuel Cycle Facilities, IAEA Safety Standards Series No. SSR-4, IAEA, Vienna (2017).

[15] INTERNATIONAL ATOMIC ENERGY AGENCY, Disposal of Radioactive Waste, IAEA Safety Standards Series No. SSR-5, IAEA, Vienna (2011).

[16] INTERNATIONAL ATOMIC ENERGY AGENCY, Borehole Disposal Facilities for Radioactive Waste, IAEA Safety Standards Series No. SSG-1, IAEA, Vienna (2009).

[17] INTERNATIONAL ATOMIC ENERGY AGENCY, Geological Disposal Facilities for Radioactive Waste, IAEA Safety Standards Series No. SSG-14, IAEA, Vienna (2011).

[18] INTERNATIONAL ATOMIC ENERGY AGENCY, Storage of Spent Nuclear Fuel, IAEA Safety Standards Series No. SSG-15, IAEA, Vienna (2012).

[19] INTERNATIONAL ATOMIC ENERGY AGENCY, Near Surface Disposal Facilities for Radioactive Waste, IAEA Safety Standards Series No. SSG-29, IAEA, Vienna (2014).

[20] INTERNATIONAL ATOMIC ENERGY AGENCY, Predisposal Management of Radioactive Waste from Nuclear Power Plants and Research Reactors, IAEA Safety Standards Series No. SSG-40, IAEA, Vienna (2016).

[21] INTERNATIONAL ATOMIC ENERGY AGENCY, Predisposal Management of Radioactive Waste from Nuclear Fuel Cycle Facilities, IAEA Safety Standards Series No. SSG-41, IAEA, Vienna (2016).

[22] INTERNATIONAL ATOMIC ENERGY AGENCY, Safety of Nuclear Fuel Reprocessing Facilities, IAEA Safety Standards Series No. SSG-42, IAEA, Vienna (2017).

[23] INTERNATIONAL ATOMIC ENERGY AGENCY, Predisposal Management of Radioactive Waste from the Use of Radioactive Material in Medicine, Industry, Agriculture, Research and Education, IAEA Safety Standards Series No. SSG-45, IAEA, Vienna (2019).

[24] INTERNATIONAL ATOMIC ENERGY AGENCY, Storage of Radioactive Waste, IAEA Safety Standards Series No. WS-G-6.1, IAEA, Vienna (2006).

[25] INTERNATIONAL ATOMIC ENERGY AGENCY, Policies and Strategies for Radioactive Waste Management, IAEA Nuclear Energy Series No. NW-G-1.1, IAEA, Vienna (2009).

[26] INTERNATIONAL ATOMIC ENERGY AGENCY, Options for Management of Spent Nuclear Fuel and Radioactive Waste for Countries Developing New Nuclear Power Programmes, IAEA Nuclear Energy Series No. NW-T-1.24 (Rev. 1), IAEA, Vienna (2018).

[27] INTERNATIONAL ATOMIC ENERGY AGENCY, Management of Disused Sealed Radioactive Sources, IAEA Nuclear Energy Series No. NW-T-1.3, IAEA, Vienna (2014).

[28] INTERNATIONAL ATOMIC ENERGY AGENCY, Framework and Challenges for Initiating Multinational Cooperation for the Development of a Radioactive Waste Repository, IAEA Nuclear Energy Series No. NW-T-1.5, IAEA, Vienna (2016).

[29] INTERNATIONAL ATOMIC ENERGY AGENCY, Storing Spent Fuel Until Transport to Reprocessing or Disposal, IAEA Nuclear Energy Series No. NF-T-3.3, IAEA, Vienna (2019).

[30] INTERNATIONAL ATOMIC ENERGY AGENCY, Stakeholder Involvement Throughout the Life Cycle of Nuclear Facilities, IAEA Nuclear Energy Series No. NG-T-1.4, IAEA, Vienna (2011).

[31] INTERNATIONAL ATOMIC ENERGY AGENCY, Available Reprocessing and Recycling Services for Research Reactor Spent Nuclear Fuel, IAEA Nuclear Energy Series No. NW-T-1.11, IAEA, Vienna (2017).

[32] INTERNATIONAL ATOMIC ENERGY AGENCY, An Overview of Stakeholder Involvement in Decommissioning, IAEA Nuclear Energy Series N0. NW-T-2.5, IAEA, Vienna (2009).

[33] INTERNATIONAL ATOMIC ENERGY AGENCY, Locating and Characterizing Disused Sealed Radioactive Sources in Historical Waste, IAEA Nuclear Energy Series No. NW-T-1.17, IAEA, Vienna (2009).

[34] INTERNATIONAL ATOMIC ENERGY AGENCY, Experiences and Lessons Learned Worldwide in the Cleanup and Decommissioning of Nuclear Facilities in the Aftermath of Accidents, IAEA Nuclear Energy Series No. NW-T-2.7, IAEA, Vienna (2014).

[35] INTERNATIONAL ATOMIC ENERGY AGENCY, Communication and Stakeholder Involvement in Environmental Remediation Projects, IAEA Nuclear Energy Series No. NW-T-3.5, IAEA, Vienna (2019).

[36] INTERNATIONAL ATOMIC ENERGY AGENCY, Nuclear Power Reactors in the World, IAEA Reference Data Series No. 2, IAEA, Vienna (2017).

[37] INTERNATIONAL ATOMIC ENERGY AGENCY, Fundamental Safety Principles, IAEA Safety Standards Series No. SF-1, IAEA, Vienna (2006).

[38] INTERNATIONAL ATOMIC ENERGY AGENCY, Governmental, Legal and Regulatory Framework for Safety, IAEA Safety Standards Series No. GSR Part 1 (Rev. 1), IAEA, Vienna (2016).

[39] OECD NUCLEAR ENERGY AGENCY, The Evolving Role and Image of the Regulator in Radioactive Waste Management: Trends over Two Decades, OECD, Paris (2012).

[40] EUROPEAN COMMISSION, Seventh Situation Report on Radioactive Waste and Spent Fuel Management in the European Union, SEC(2011), European Commission, Brussels (2011).

[41] WORLD NUCLEAR ASSOCIATION, Nuclear Power Economics and Project Structuring, WNA, London (2017).

[42] OECD NUCLEAR ENERGY AGENCY, The Economics of the Back End of the Nuclear Fuel Cycle, NEA Report No. 7061, OECD, Paris (2013).

[43] OECD NUCLEAR ENERGY AGENCY, Low-level Radioactive Waste Repositories, An Analysis of Costs, OECD, Paris (1999).

[44] SWEDISH NUCLEAR FUEL AND WASTE MANAGEMENT CO, SKB, Plan 2016. Costs From and Including 2018 for the Radioactive Residual Products from Nuclear Power, TR-17-02, SKB, Stockholm (2017).

[45] INTERNATIONAL ATOMIC ENERGY AGENCY, IAEA Safety Glossary: 2018 Edition, IAEA, Vienna (2019).

[46] INTERNATIONAL NUCLEAR SAFETY GROUP, Stakeholder Involvement in Nuclear Issues, IAEA INSAG series No. 20, IAEA, Vienna (2006).

[47] INTERNATIONAL ATOMIC ENERGY AGENCY, Communications on Nuclear, Radiation, Transport and Waste Safety: A Practical Handbook, IAEA-TECDOC-1076, IAEA, Vienna (1999).

[48] INTERNATIONAL ATOMIC ENERGY AGENCY, Factors Affecting Public and Political Acceptance for the Implementation of Geological Disposal, IAEA-TECDOC-1566, IAEA, Vienna (2007).

[49] UNITED NATIONS ECONOMIC COMMISSION FOR EUROPE, Convention on Access to Information, Public Participation in Decision-Making and Access to Justice in Environmental Matters, UNECE, Geneva (1998).

[50] UNITED NATIONS ECONOMIC COMMISSION FOR EUROPE, Convention on Environmental Impact Assessment in a Transboundary Context, UNECE, Geneva (1991).

[51] NUCLEAR DECOMMISSIONING AUTHORITY, Strategy, Effective from April 2016. SG/2016/53, NDA, Moor Row, UK (2016).

[52] INTERNATIONAL ATOMIC ENERGY AGENCY, Pressurized Heavy Water Reactor Fuel: Integrity, Performance and Advanced Concepts, IAEA-TECDOC-CD-1751, IAEA, Vienna (2014).

[53] INTERNATIONAL ATOMIC ENERGY AGENCY, Status and Advances in MOX Fuel Technology, IAEA Technical Reports Series No. 415, IAEA, Vienna (2003).

[54] UNITED STATES DEPARTMENT OF ENERGY, A Historical Review of the Safe Transport of Spent Nuclear Fuel, Prepared for US Department of Energy Nuclear Fuels Storage and Transportation Planning Project, USDOE, Washington, DC (2016).

[55] INTERNATIONAL ATOMIC ENERGY AGENCY, Operation and Maintenance of Spent Fuel Storage and Transportation Casks/Containers, IAEA-TECDOC-1532, IAEA, Vienna (2007).

[56] INTERNATIONAL ATOMIC ENERGY AGENCY, Regulations for the Safe Transport of Radioactive Material, IAEA Safety Standards Series No. SSR-6 (Rev. 1), IAEA, Vienna (2018).

[57] INTERNATIONAL ATOMIC ENERGY AGENCY, Selection of Away-from-Reactor Facilities for Spent Fuel Storage, IAEA-TECDOC-CD-1558, IAEA, Vienna (2007).

[58] OECD NUCLEAR ENERGY AGENCY, The Safety of Long-Term Interim Storage Facilities in NEA Member Countries, OECD, Paris (2017).

[59] INTERNATIONAL ATOMIC ENERGY AGENCY, Scientific and Technical Basis for the Geological Disposal of Radioactive Wastes, IAEA Technical Reports Series No. 413, IAEA, Vienna (2003).

[60] NATIONAL RESEARCH COUNCIL, Disposition of High-Level Waste and Spent Nuclear Fuel: The Continuing Societal and Technical Challenges, The National Academies Press, Washington, DC (2001).

[61] OECD NUCLEAR ENERGY AGENCY, Int. Conf. on Geological Repositories 2016. Conf. Synthesis 7–9 December 2016, NEA No. 7345, OECD, Paris (2017).

[62] INTERNATIONAL ATOMIC ENERGY AGENCY, The Use of Scientific and Technical Results from Underground Research Laboratory Investigations for the Geological Disposal of Radioactive Waste, IAEA-TECDOC-1243, IAEA, Vienna (2001).

[63] OECD NUCLEAR ENERGY AGENCY, Underground Research Laboratories (URL), No. 78122, OECD, Paris (2013).

[64] INTERNATIONAL ATOMIC ENERGY AGENCY, Application of Thermal Technologies for Processing of Radioactive Waste, IAEA-TECDOC-1527, IAEA, Vienna (2006).

[65] INTERNATIONAL ATOMIC ENERGY AGENCY, Selection of Technical Solutions for the Management of Radioactive Waste, IAEA-TECDOC-1817, IAEA, Vienna (2017).

[66] NATIONAL ACADEMIES OF SCIENCES, ENGINEERING, AND MEDICINE, Low-Level Radioactive Waste Management and Disposition: Proc. Workshop, The National Academies Press, Washington, DC (2017).

[67] EUROPEAN UNION, Council Directive 2013/59/Euratom of 5 December 2013 laying down basic safety standards for protection against the dangers arising from exposure to ionising radiation, and repealing Directives 89/618/Euratom, 90/641/Euratom, 96/29/Euratom, OJEU L13 (2014).

[68] EUROPEAN UNION, Council Regulation 1493/93/Euratom of 8 June 1993 on shipments of radioactive substances between Member States, OJEC L148 (1993).

[69] EUROPEAN NUCLEAR SAFETY REGULATOR'S GROUP WG2, Guidelines for Member States reporting on Article 14.1 of Council Directive 2011/70/Euratom, ENSREG (2018).

[70] OECD NUCLEAR ENERGY AGENCY, National Inventories and Management Strategies for Spent Nuclear Fuel and Radioactive Waste. Methodology for Common Presentation of Data, OECD, Paris (2016).

[71] NATIONAL ACADEMIES OF SCIENCES, ENGINEERING, AND MEDICINE, Reducing the Use of Highly Enriched Uranium in Civilian Research Reactors, The National Academies Press, Washington, DC (2016).

[72] INTERNATIONAL ATOMIC ENERGY AGENCY, Further Analysis of Extended Storage of Spent Fuel, IAEA-TECDOC-944, IAEA, Vienna (1997).

[73] INTERNATIONAL ATOMIC ENERGY AGENCY, Behaviour of Spent Power Reactor Fuel During Storage, IAEA-TECDOC-1862, IAEA, Vienna (2019).

[74] UNITED STATES DEPARTMENT OF ENERGY, United States of America Sixth National Report for the Joint Convention on the Safety of Spent Fuel Management and on the Safety of Radioactive Waste Management, USDOE, Washington, DC (2017).

[75] UNITED STATES DEPARTMENT OF ENERGY, EIS-0375: Disposal of Greater-than-Class-C Low-Level Radioactive Waste and Department of Energy GTCC-like Waste, USDOE, Washington, DC (2016).

[76] STATE ATOMIC ENERGY CORPORATION "ROSATOM", The Fifth National Report of the Russian Federation on Compliance with the Obligations of the Joint Convention on the Safety of Spent Fuel Management and the Safety of Radioactive Waste Management, Rosatom, Moscow (2017).

[77] OECD NUCLEAR ENERGY AGENCY, The Management of Large Components from Decommissioning to Storage and Disposal, NEA/RWM/R(2012)8, OECD, Paris (2012).

[78] INTERNATIONAL ATOMIC ENERGY AGENCY, Chernobyl: Looking Back to Go Forward, Proc. Int. Conf. held in Vienna, Austria, 6–7 September 2005, IAEA, Vienna (2008).

[79] INTERNATIONAL ATOMIC ENERGY AGENCY, The Fukushima Daiichi Accident, IAEA, Vienna (2015).

[80] INTERNATIONAL ATOMIC ENERGY AGENCY, Mission Report IAEA International Peer Review Mission on Mid- and Long-Term Roadmap Towards the Decommissioning of TEPCO's Fukushima Daiichi Nuclear Power Station, IAEA, Vienna (2018).

[81] INTERNATIONAL ATOMIC ENERGY AGENCY, Remediation Process for Areas Affected by Past Activities and Accidents, IAEA Safety Standards Series No. WS-G-3.1, IAEA, Vienna (2007).

[82] INTERNATIONAL ATOMIC ENERGY AGENCY, Demonstrating Performance of Spent Fuel and Related Storage System Components during Very Long Term Storage, IAEA-TECDOC-1878, IAEA, Vienna (2019).

[83] INTERNATIONAL ATOMIC ENERGY AGENCY, Spent Fuel Performance Assessment and Research: Final Report of a Coordinated Research Project (SPAR-II), IAEA-TECDOC-1680, IAEA, Vienna (2012).

[84] INTERNATIONAL ATOMIC ENERGY AGENCY, Spent Fuel Performance Assessment and Research: Final Report of a Coordinated Research Project on Spent Fuel Performance Assessment and Research (SPAR-III) 2009–2014, IAEA-TECDOC-1771, IAEA, Vienna (2015).

[85] UNITED STATES DEPARTMENT OF ENERGY, EPRI/DOE High-Burnup Fuel Sister Rod Test Plan Simplification and Visualization, USDOE, Washington, DC (2017).

[86] INTERNATIONAL ATOMIC ENERGY AGENCY, Fifth Review Meeting of the Contracting Parties to the Joint Convention on the Safety of Spent Fuel Management and on the Safety of Radioactive Waste Management, Final Summary Report, JC/RM5/04/Rev. 2, IAEA, Vienna (2015).

[87] INTERNATIONAL ATOMIC ENERGY AGENCY, BOSS: Borehole Disposal of Disused Sealed Sources, IAEA-TECDOC-1644, IAEA, Vienna (2011).

[88] INTERNATIONAL ATOMIC ENERGY AGENCY, Enhancing Benefits of Nuclear Energy Technology Innovation through Cooperation among Countries: Final Report of the INPRO Collaborative Project SYNERGIES, IAEA Nuclear Energy Series No. NF-T-4.9, IAEA, Vienna (2018).

[89] OECD NUCLEAR ENERGY AGENCY, Cost of Decommissioning Nuclear Power Plants, OECD, Paris (2016).

[90] INTERNATIONAL ATOMIC ENERGY AGENCY, Data Analysis and Collection for Costing of Research Reactor Decommissioning, IAEA-TECDOC-1832, IAEA, Vienna (2017).

CONTENTS OF THE ANNEXES

The on-line supplementary files for this publication can be found on the publication's web page at https://www.iaea.org/publications.

ABBREVIATIONS

AFR	away-from-reactor
CRP	coordinated research project
DGR	deep geological repository
DSRS	disused sealed radioactive source
DU	depleted uranium
EGIRM	Expert Group on Waste Inventorying and Reporting Methodology
Euratom	European Atomic Energy Community
EU	European Union
HLW	high level waste
ILW	intermediate level waste
LLW	low level waste
LWR	light water reactor
MOX	mixed oxide
NORM	naturally occurring radioactive material
NPP	nuclear power plant
OECD/NEA	OECD Nuclear Energy Agency
OECD	Organisation for Economic Co-operation and Development
R&D	research and development
t HM	tonnes of heavy metal
UMM	uranium mining and milling
URL	underground research laboratory
VLLW	very low level waste
WIPP	Waste Isolation Pilot Plant
WMO	waste management organization
WNA	World Nuclear Association

CONTRIBUTORS TO DRAFTING AND REVIEW

Alvarez, Y.R.	National Nuclear Safety and Safeguards Commission, Mexico
Babilas, E.	Lithuanian Energy Institute, Lithuania
Baila, S.	Swissnuclear, Switzerland
Batandjieva-Metcalf, B.	European Commission
Bevilacqua, A.	National Atomic Energy Commission, Argentina
Canak, M.	Fund for financing the decommissioning of the Krško NPP, Croatia
Caruso, S.	National Cooperative for the Disposal of Radioactive Waste, Switzerland
Chepurnyi, I.	State Nuclear Regulatory Inspectorate of Ukraine, Ukraine
Deryabin, S.	State Atomic Energy Corporation Rosatom, Russian Federation
Dima, M.A.	Nuclear and Radioactive Waste Agency, Romania
Dionisi, M.	Italian National Institute for Environmental Protection and Research, Italy
Dutzer, M.	Consultant
Forsström, H.	Consultant
Garamszeghy, M.	Consultant
Garcia Neri, E.	Enresa, Spain
Gastl, C.	International Atomic Energy Agency
González-Espartero, A.	International Atomic Energy Agency
Gordon, I.A.	International Atomic Energy Agency
Grzegrzółka, G.	Radioactive Waste Management Plant, Poland
Guiot, B.	Federal Agency for Nuclear Control, Belgium
Gunter, T.	Department of Energy, United States of America
Hassan, A.	Iran Nuclear Regulatory Authority, Islamic Republic of Iran
Heath, M.	Nuclear Regulatory Commission, United States of America
Hedberg, B.	Swedish Radiation Safety Authority, Sweden
Henderson, D.	OECD Nuclear Energy Agency
Ivanoviç, T.	Nuclear Regulatory Authority of the Slovak Republic, Slovakia
James, M.	Nuclear Decommissioning Authority, United Kingdom
Janušytė, I.	Radioactive Waste Management Agency, Lithuania
Kenny, J.	Natural Resources Canada, Canada

Kozich, H.	Low Level Waste Repository, United Kingdom
Kuciel, G	Radioactive Waste Management Plant, Poland
Lazar, I.	Hungarian Atomic Energy Authority, Hungary
Lebedev, V.	OECD Nuclear Energy Agency
Lemmens, A.	Belgian Agency for Radioactive Waste and Enriched Fissile Materials, Belgium
Li, F.	China Atomic Energy Authority, China
Lust, M.	International Atomic Energy Agency
Maital, M.	Moroccan Agency for Nuclear and Radiological Safety and Security, Morocco
Matuzas, V.	European Commission
Mekki, S.	French National Radioactive Waste Management Agency, France
Metaxopoulou, I.	European Commission
Mohamed, Y.S.	Egyptian Atomic Energy Authority, Egypt
Ognerubov, V.	Ignalina Nuclear Power Plant, Lithuania
O'Sullivan, P.J.	International Atomic Energy Agency
Paterson, H.C.	Nuclear Decommissioning Authority, United Kingdom
Petrosyan, A.	Ministry of Energy Infrastructures and Natural Resources, Armenia
Petry, E.	French National Radioactive Waste Management Agency, France
Pieraccini, M.	Électricité de France, France
Place, I.	Orano, France
Pölcher, U.	European Commission
Rahayu, D.	National Nuclear Energy Agency, Indonesia
Robbins, R.	International Atomic Energy Agency
Rothwell, G.	OECD Nuclear Energy Agency
Saghiu, V.I.	Nuclear and Radioactive Waste Agency, Romania
Šatrovska, D.	Radiation Safety Centre, Latvia
Schmidt, M.	Federal Office for Radiation Protection, Germany
Shenk, J.	Department of Energy, United States of America
Stark, K.	Swedish Radiation Safety Authority, Sweden
Strässle, C.	Decommissioning Fund for Nuclear Facilities and Waste Disposal Fund for Nuclear Power Plants, Switzerland
Sundin, S.	Swedish Radiation Safety Authority, Sweden

Tomar, N.	Bhabha Atomic Research Centre, India
Tonkay, D.	Department of Energy, United States of America
Turner, M.	Nuclear Regulatory Authority of the Slovak Republic, Slovakia
Twala, V.G.	National Radioactive Waste Disposal Institute, South Africa
Ünver, L.O.	Turkish Atomic Energy Authority, Turkey
Vela-García, M.	European Commission
Viorel, T.L.	Nuclear and Radioactive Waste Agency, Romania
Wahab Yusof, M.A.	Malaysian Nuclear Agency, Malaysia
Xerri, C.	International Atomic Energy Agency
Yuen, P.	Natural Resources Canada, Canada
Zaccai, H.	World Nuclear Association
Zherebtsov, A.	State Atomic Energy Corporation Rosatom, Russian Federation
Zhurbenko, E.	TENEX, Russian Federation

Technical Meetings

Paris, France: 12–16 June 2017

Luxembourg: 2–6 July 2018

Vienna, Austria: 9–11 July 2019

Consultants Meetings

Paris, France: 12–16 June 2017

Vienna, Austria: 29 January–2 February 2018, 5–7 February 2019, 9–11 July 2019

Luxembourg: 2–6 July 2018

Structure of the IAEA Nuclear Energy Series*

Nuclear Energy Basic Principles
NE-BP

Nuclear Energy General Objectives
NG-O

1. Management Systems
NG-G-1.#
NG-T-1.#

2. Human Resources
NG-G-2.#
NG-T-2.#

3. Nuclear Infrastructure and Planning
NG-G-3.#
NG-T-3.#

4. Economics and Energy System Analysis
NG-G-4.#
NG-T-4.#

5. Stakeholder Involvement
NG-G-5.#
NG-T-5.#

6. Knowledge Management
NG-G-6.#
NG-T-6.#

Nuclear Reactor** Objectives
NR-O

1. Technology Development
NR-G-1.#
NR-T-1.#

2. Design, Construction and Commissioning of Nuclear Power Plants
NR-G-2.#
NR-T-2.#

3. Operation of Nuclear Power Plants
NR-G-3.#
NR-T-3.#

4. Non Electrical Applications
NR-G-4.#
NR-T-4.#

5. Research Reactors
NR-G-5.#
NR-T-5.#

Nuclear Fuel Cycle Objectives
NF-O

1. Exploration and Production of Raw Materials for Nuclear Energy
NF-G-1.#
NF-T-1.#

2. Fuel Engineering and Performance
NF-G-2.#
NF-T-2.#

3. Spent Fuel Management
NF-G-3.#
NF-T-3.#

4. Fuel Cycle Options
NF-G-4.#
NF-T-4.#

5. Nuclear Fuel Cycle Facilities
NF-G-5.#
NF-T-5.#

Radioactive Waste Management and Decommissioning Objectives
NW-O

1. Radioactive Waste Management
NW-G-1.#
NW-T-1.#

2. Decommissioning of Nuclear Facilities
NW-G-2.#
NW-T-2.#

3. Environmental Remediation
NW-G-3.#
NW-T-3.#

(*) as of 1 January 2020
(**) Formerly 'Nuclear Power' (NP)

Key
BP: Basic Principles
O: Objectives
G: Guides and Methodologies
T: Technical Reports
Nos 1–6: Topic designations
#: Guide or Report number

Examples
NG-G-3.1: Nuclear Energy General (**NG**), Guides and Methodologies (**G**), Nuclear Infrastructure and Planning (topic **3**), **#1**
NR-T-5.4: Nuclear Reactors (**NR**)*, Technical Report (**T**), Research Reactors (topic **5**), **#4**
NF-T-3.6: Nuclear Fuel (**NF**), Technical Report (**T**), Spent Fuel Management (topic **3**), **#6**
NW-G-1.1: Radioactive Waste Management and Decommissioning (**NW**), Guides and Methodologies (**G**), Radioactive Waste Management (topic **1**) **#1**

IAEA
International Atomic Energy Agency

ORDERING LOCALLY

IAEA priced publications may be purchased from the sources listed below or from major local booksellers.

Orders for unpriced publications should be made directly to the IAEA. The contact details are given at the end of this list.

NORTH AMERICA

Bernan / Rowman & Littlefield

15250 NBN Way, Blue Ridge Summit, PA 17214, USA
Telephone: +1 800 462 6420 • Fax: +1 800 338 4550

Email: orders@rowman.com • Web site: www.rowman.com/bernan

REST OF WORLD

Please contact your preferred local supplier, or our lead distributor:

Eurospan Group

Gray's Inn House
127 Clerkenwell Road
London EC1R 5DB
United Kingdom

Trade orders and enquiries:

Telephone: +44 (0)176 760 4972 • Fax: +44 (0)176 760 1640
Email: eurospan@turpin-distribution.com

Individual orders:

www.eurospanbookstore.com/iaea

For further information:

Telephone: +44 (0)207 240 0856 • Fax: +44 (0)207 379 0609
Email: info@eurospangroup.com • Web site: www.eurospangroup.com

Orders for both priced and unpriced publications may be addressed directly to:

Marketing and Sales Unit
International Atomic Energy Agency
Vienna International Centre, PO Box 100, 1400 Vienna, Austria
Telephone: +43 1 2600 22529 or 22530 • Fax: +43 1 26007 22529
Email: sales.publications@iaea.org • Web site: www.iaea.org/publications